D0296415

A Mothers Day Gift

- so we can all play a part

as a family to save our planet.

Much love,

Mum & Dad.

HOW TO

SAVE YOUR PLANET

ONE

OBJECT

AT A TIME

First published in Great Britain by Simon & Schuster UK Ltd, 2020

A CBS COMPANY

1 3 5 7 9 10 8 6 4 2

Simon & Schuster UK Ltd
1st Floor, 222 Gray's Inn Road
London WC1X 8HB

www.simonandschuster.co.uk
www.simonandschuster.com.au
www.simonandschuster.co.in

Simon & Schuster Australia, Sydney
Simon & Schuster India, New Delhi

A CIP catalogue record for this book is available from the British Library

Hardback ISBN: 978-1-4711-8410-9

Ebook ISBN: 978-1-4711-8411-6

Illustrations:
Pages 12, 16–17, 20–21, 24–25, 30–31, 34–35, 36–37, 38–39, 50, 52–53, 55, 56–57, 70, 79–81, 84–85, 86, 90, 95, 97, 100, 102–104, 108, 110, 116, 118, 124, 126, 130, 132–133, 142–143, 144–145, 147–148, 152, 156–157, 158, 159–160, 161, 162–163, 166–167, 168–169, 178, 179, 180, 182–183, 188, 190–191, 200, 202–203, 208–209, 214–215, 218–219, 224–225, 226, 234–235, 237, 240–241, 244–245 © Tonwen Jones

Pages 8, 13, 14–15, 26–27, 28–29, 32–33, 40, 42–43, 44, 51, 58–59, 64–65, 66, 68, 71, 72, 76, 82, 92, 94, 98–99, 101, 105–107, 114–115, 117, 119, 120–122, 128, 136–138, 140–141, 149, 151, 153, 154–155, 164, 171, 172, 174, 176–177, 181, 184, 186, 193, 195, 196–197, 198, 201, 204, 206, 211, 213, 216–217, 220–221, 222, 227, 228–229, 231–232, 236, 238–239, 242–243, 247 © Nicholas Stevenson

Photograph page 256 © Leon Farrell/Photocall Ireland

Editorial Director: Fritha Saunders
Project Editor: Nikki Sims
Design: Rachel Cross
Cover Design: Siân Wilson
Production: Karin Seifried

Typeset in Din, Georgia and Univers.

Printed in Italy with vegetable-based inks.

DR TARA SHINE

SIMON &
SCHUSTER

London · New York · Sydney · Toronto · New Delhi

A CBS COMPANY

CONTENTS

FOREWORD

Dr Tara Shine is an enlightened, big-picture thinker, and with this book she shows that she is equally and delightfully adept at bringing details into focus.

I first met Tara in the context of the international climate change negotiations that led to the adoption of the Paris Agreement. I was leading the work of the United Nations Secretariat of the Convention on Climate Change and Tara was in the trenches working for a legally binding agreement shaped by fairness and justice. Then, as now, she was determined and steadfast, driven by her principles and her desire to find practical solutions to climate change that would benefit the many and not the few. At the time, hoping for a new agreement was ambition enough for many people, ensuring it would be fair and inclusive was seen as wildly optimistic. Luckily Tara shares my unrelenting optimism on the common good, and she worked alongside others from indigenous women to small-scale farmers to bring justice into the discussions. Working alongside President Mary Robinson, Tara was part of the effort to ensure that women's voices and leadership were harnessed in support of what became a global legally binding accord to protect the future of humanity and the planet.

Many of us who work at the international level on global problems such as climate change are asked about the value of individual actions when the imperative is to change the behaviour of states and corporations. The truth of the matter is that we have the international framework in place and what is needed now is action at country and local level. Tara recognises this action depends on people and how they are engaged and feel empowered to be part of positive change. Giving people the information they need to make informed choices is an important step in helping the national conversation and national commitments to grow.

This book is all about realising the power you have as an individual by informing yourself, asking questions and making smart choices. By becoming active and joining the conversation, you become empowered and you do something about the problem we face rather than feeling powerless in its presence.

Tara uses her knowledge and expertise as an environmental scientist and policy adviser to bring robust and insightful information to everyday objects, tasks and habits in an easy-to-read way. She mixes practical advice with scientific data to bring the latest information on how to live sustainably into your life, your home and your office. She also brings the latest innovations to our attention alongside the wisdom and frugal good sense of our grandparents. Puncturing the myth that sustainability is a complex and unattainable goal, Tara puts sustainability literally at our fingertips. I am constantly heartened by the action individuals, businesses and communities are taking around the world as they innovate to reduce emissions, reduce waste and become more resilient. From cities to schools amazing leadership is being shown in the race to make our world more sustainable.

This book equips activists in bedrooms and boardrooms to lead by example, to show that change can be a desirable thing and that there is a different, kinder and lighter way to live on planet Earth. I recommend this book to anyone who wants to do the right thing but isn't quite sure where to start, or to someone who has started their journey to more sustainable living and wants to know more and do more. And for your further involvement I can only recommend Tara's social enterprise Change by Degrees, www.changebydegrees.com.

I am a stubborn climate optimist and that trait allows me to doggedly pursue a vision of a better world. I invite you to be a climate optimist, too, and to use what you learn in this book to equip yourself to be a force for change. If you are a consumer, a voter, a volunteer, a worker or a parent, read this book and learn what you can do today that will make a difference. Our collective actions can fuel a shared conversation for change that is both powerful and necessary. Above all it is urgent.

Christiana Figueres

Former Executive Secretary of the United Nations Framework Convention on Climate Change (UNFCCC)

Convenor Mission 2020

Founding partner of Global Optimism Ltd

Change: To make or become different.
A new or refreshingly different experience.

INTRODUCTION

This is a book about change, with the purpose of making things better than they are now. It is filled with positivity and optimism about the future, belief in humanity and excitement about our powers of innovation. It focuses on everyday objects and the habits, rituals and behaviours associated with them, all of which we can change to reduce the impact we have on our environment – plus they'll improve our health and wellbeing. It is also an honest book, about the scale of the existential crises facing humanity, caused by the way we live on and use the resources of planet Earth. In the face of a crisis or an emergency, decisive action is required by governments, businesses and people. The most important thing to do is to act, to start to do something. And this book is filled with what you can do in your day-to-day life to reduce your impact on our planetary home.

ALL TOGETHER NOW

It can be argued that in the face of global problems individual action is inconsequential. But this misses the point. Climate change, plastic pollution and biodiversity loss are caused by the individual and collective actions of the 7.7 billion of us who live on Earth. So, the answer lies in us and what we do, what we buy, how we act and what we value. Of course, I am not claiming that everyone giving up cling film will save the planet, but becoming informed, having and sharing an opinion and changing what we buy and how we live does matter. We also need supporting international and national policies and laws – but they are also created by people and need the support of citizens to be passed through parliaments, funded and implemented on the ground. The question is not whether we need top-down or bottom-up solutions – we need both.

For twenty years I worked at an international level advising governments on policies and legal agreements to enable countries to move to a more sustainable development pathway. Along the way there were huge frustrations – like the failed climate summit in Copenhagen in 2009 – as well as the high points, such as the celebrations that accompanied the adoption of the Paris Agreement on climate change in 2015. I have seen the power of multilateral policy-making in shaping agendas to make our world fairer, safer and healthier and I have seen the influence those agreements have when translated into policy and law at a national level. But I have also seen good policies fail at national level because they didn't have the support of the people or because politicians feared losing votes due to unpopular policy choices. So, in the last few years I have turned my attention to the people – you and me – because without our engagement the progressive policy changes we need may not happen.

Today I help people at home and at work to make small changes to everyday things to reduce their environmental impact and add to the quality of their life. Starting to make changes in our own lives is extremely motivating and helps us to realise the power we have. We are more than private individuals; we are voters, consumers, shareholders, volunteers, activists, stakeholders and communities and our influence is significant.

I believe that in order to make the systemic changes we need to avoid a planetary extinction, we need to engage people in the conversation about what that change should be and how it can be better than what we have now. This book aims to play its part in changing the conversation so that we are ready to vote for and be part of the systemic change we so urgently need.

The changes and actions proposed in this book do not require massive investments of your time or money – many are simply changes of habit that will quite likely also improve your quality of life. You can start by choosing one or two changes to make based on the ideas in this book and if that is easily manageable, move on to another and then another.

My hope is that you will see the power of your actions, spread the word and see the value of being part of a growing movement of people who want to find a better way of living on planet Earth. Harder choices follow, such as taking fewer flights for holidays or work, installing solar panels on your roof or living without a car, but if your starting place is giving up disposable coffee cups, then begin there.

LESS STUFF IS GOOD

A core message of this book is that the most sustainable objects are often the ones you already own. Although I suggest many sustainable alternatives, this is not intended as an invitation to throw everything out and start again by buying new, more sustainable stuff! Most of us already have too much stuff and this book is ultimately about owning less and using and reusing what we already have more wisely. At times I refer to specific brands or products – these are not endorsements but are included to tell a story or show how change is possible.

Habit and ritual are other core themes explored in this book. Most habits are hard to break, but some of the habits we have formed recently, in the last few decades, are threatening our very existence. The concept of a takeaway coffee in a disposable cup didn't exist when my father was a young man and the modern-day wet wipe was not in common use when I was a baby. These are all new habits we have acquired in a relatively short space of time.

We know from experience that new habits can be adopted in a surprisingly short space of time when: the rationale for doing so is clear; alternatives are available (such as reusable shopping bags);

and new social norms developed to support the change – just look how quickly we got used to wearing seatbelts in cars and not smoking in many public spaces.

I have drawn on my extensive experience to bring you the most robust information I can find in an accessible way. I have gained insights from my time as a negotiator of international climate and sustainable development policy, as a researcher in Europe and in developing countries and as an adviser to world leaders. I am honest about the scale of the challenges facing humanity because the science supporting these assessments is emphatic, but I want you to read this book and know that there are things you can do in the face of these challenges that matter and will make a difference. That is why this book is intensely practical and proposes actions that should resonate in daily life. I am also a mum, a worker, a volunteer and a daughter – so I know what real life is like with all of the challenges and opportunities it throws up. Of course, the science continues to develop and innovation is happening all the time, so bear in mind that the information in this book is based on the best available information at the time of writing and may at some point be subject to change.

SUSTAINABILITY NUTS AND BOLTS

Sustainability means a healthy planet, a thriving economy and a happy society that uses the Earth's resources in a way that respects the needs of the people and biodiversity we share the planet with today and preserves and protects them for future generations.

In 2015, all of the nations of the world came together under the United Nations and adopted the seventeen Sustainable Development Goals (SDGs). The goals set an inspirational and ambitious agenda for a better world by 2030. They cover everything from health and education to food waste and climate change and every country in the world has committed to working to achieve them.

WHY WE SHOULD AIM FOR 1.5°C

Research conducted by the Intergovernmental Panel on Climate Change (IPCC) and published in 2018 revealed that a rise of 2°C is not a safe level of warming and that aiming for 1.5°C is a global imperative.

- At 1.5°C of warming, 6 per cent of insects, 8 per cent of plants and 4 per cent of vertebrates (animals with a backbone) will be lost, while at 2°C this rises to 18 per cent of insects, 16 per cent of plants and 8 per cent of vertebrates. With this loss of biodiversity, we lose food crops and species that could hold the cure to diseases.
- At 1.5°C the probability of an ice-free Arctic Ocean during the summer is once per century; at 2°C it is once per decade. This will affect the livelihoods of the indigenous people who live in the Arctic as well as the survival chances of animals that live there, such as polar bears.

In order to achieve the 1.5°C goal we need to reduce the greenhouse gas emissions we produce to zero by 2050 at the latest. As it stands, our greenhouse gas emissions keep growing. Instead, we – every person, family, school, business, town and country – need to halve these emissions every decade to keep the Earth's climate stable. That means more insulation, renewable energy, public transport, tree planting and a dramatic reduction in how much stuff we buy and throw away.

Also in 2015, The Paris Agreement, a legally binding accord, was agreed. It aims to reduce greenhouse gas emissions, encourage investment in a low-carbon future and help countries adapt to the inevitable impacts of climate change. The agreement also set important temperature goals, committing all countries in the world to limiting global temperature rise to less than 2°C above pre-industrial levels (that is, when the world started to industrialise and use fossil fuels) and to aim for the safer limit of 1.5°C. Countries have submitted plans detailing how they will play their part implementing the agreement; they revise the plans upwards every five years, with the next revision in 2020. Every country in the world, bar the US which withdrew in June 2017, has signed up to the accord and is part of a global effort to achieve these goals.

FINDING THE RIGHT SOLUTIONS

Science tells us that we have until just 2030 to undertake a zero-carbon revolution that surpasses the magnitude of the Industrial Revolution and reduces global emissions by at least 45 per cent. But unlike the Industrial Revolution, we need to do it in a way that is fair, benefits all people regardless of where they live or how well off they are, in pursuit of climate justice. This means that the people who are least responsible for causing climate change, mostly those living in developing countries, need to be protected from the negative impacts of climate change and supported to reap the benefits of a low-carbon, sustainable future.

A large part of the necessary transformation will be in how we design the products we use. Any object made uses raw material and energy. So, living without an impact is impossible. However, making less stuff, using renewable energy and designing things to last and then have another use at the end of their life, can reduce the impact we have on the planet. Much of this is captured in the concept of the circular economy, which is the opposite of the take, use and dispose economic model and instead aims to create a cycle where everything is used, reused, repaired, restored and eventually returned to the Earth to create new growth. The future seen in this way is exciting and creative. It is why sustainability is about innovation and finding better ways of doing things and living.

HOW TO USE THIS BOOK

You can read this book cover to cover or dip in and out of it when you need some inspiration or information. Each object has a section on 'what you can do' that lists some actions from the easy to achieve to those that may require more effort. If you make five changes, you will be making a difference; if you make fifty you are a planetary rock star! In some cases the greenest action is to use what you have or to stop using something you could live without.

You do not have to aim for perfection. In fact, I think that perfect examples of green lifestyles, where people live in the countryside, grow their own food, produce their own energy and knit their own socks have made many of us feel that living sustainably is beyond our reach. But sustainability is something everyone can do, whether you live in a flat in the city or a house in the country, whether you are at home, do your shopping at 10p.m. in a supermarket or spend half your week commuting.

Getting started, taking action and joining the conversation gives you the power to be part of the transformation our societies need. So, which change will be your first to a more sustainable future?

THE KITCHEN

97 per cent of UK households own a kettle and more than 90 per cent of people use the kettle every day, with 40 per cent boiling the kettle five times or more a day.

KETTLE

For so many of us, the day starts by boiling the kettle to make a cup of tea or coffee. Key moments in the day are also associated with taking a break and boiling the kettle. A 2015 study looked at the contribution to global warming of 1kg of tea consumed in the UK and found that from the 'cradle to grave' it produces over 12kg of CO_2. Strikingly, 85 per cent of the impact is attributed to the electricity used to boil water.

Historically water was boiled in a pot over a fire. A kettle is a more efficient and faster way to heat water as it is pretty much a closed system, apart from the spout. The first electric kettles appeared in the 1890s, but these early kettles got very hot and could melt through the kettle casing and cause fires. The automatic switch-off-when-boiled function of a kettle wasn't invented until 1955.

Across the EU, between 117 and 200 million kettles are used every year with an estimated electricity consumption of between 19 and 33 TWh/y (terawatt hours per year). This compares to the entire annual electricity consumption of the country of Ireland, which was 26 TWh/y in 2017 and an estimated 30 TWh/y in 2018. Kettles are typically made from stainless steel or plastic; most modern kettles are cordless (no need to unplug to pour), hold 1.5–2 litres of water and are made in China.

THE IMPACTS

Most people boil more water than they use in their tea or coffee and this increases the carbon footprint of their cuppa. In fact, the most significant part of a kettle's environmental footprint is its use, so not overfilling the kettle, and not boiling it three time before you make your tea, is a significant pro-environmental action. The more water you boil, the greater the contribution to climate change.

The kettle is one of the household appliances with the highest wattage and requires the highest current when switched on. Ten years ago kettles were mostly 2.2kW appliances, but now more powerful 3kW kettles are widely available and in demand because they boil water faster – but that also means they use more electricity.

A 2016 survey of 86,000 homes in the UK, by the Energy Saving Trust, found that three-quarters of British households admit to overfilling their kettle when boiling water and are collectively paying £68 million each year more than they need to in electricity costs. The study also found that 86 per cent of people choose a kettle based on aesthetics, to match their kitchen design or another electrical appliance, rather than their energy efficiency.

WHAT YOU CAN DO

- Boil only the water you need: if you are making one cup of tea, boil one cup of water.
- Watch your kettle and boil once. A watched kettle never boils, or the saying goes, but not watching your kettle may mean you boil it multiple times to make just one cup of tea.
- Make your tea or coffee after the first boil. When replacing a kettle, choose one that allows you to see and measure the amount of water in the kettle at a glance. Also look for the most efficient A-rated kettle you can afford. Reducing the carbon footprint of your cup of tea or coffee will save you money in the long run due to lower electricity bills.
- Use a kettle to boil water rather than a microwave or a saucepan on the hob or cooker. An electric kettle converts about 80 per cent of the electricity it uses into energy to heat the water, while the comparable figure for a microwave is about 55 per cent and for a pan of water on a gas hob is about 40 per cent.
- Switch to a renewable energy provider and reduce the emissions associated with each and every cuppa.

INSTANT HOT WATER

Hot water taps may be convenient but there are mixed reviews of their environmental performance. If you are a serial overfiller and reboiler of a kettle for your cuppa, there may be a small saving from using a hot water tap. That said, it would take years to recuperate the cost of installing the tap in your home.

Additional facts

When a major event is on TV and a significant proportion of the population is watching, spikes in electricity use are anticipated as people put the kettle on at half time or during a break. In the UK, this is called 'TV pickup'. The largest ever pickup was on 4 July 1990 at the end of the penalty shootout in the England vs West Germany FIFA World Cup semi-final.

Kettle use shoots up in the UK during holidays – July, August, December, and January – when people are at home for longer periods of time.

An article in the UK's Guardian *newspaper in January 2018 stated that considering the number of cups of tea that are brewed every year in the UK, the amount of polypropylene used to seal the bags could be in the region of 150 tonnes of plastic – this accumulated waste may be contaminating food waste compost collections, going to landfill or escaping into the environment.*

TEA BAG

Whether you like strong builder's tea or prefer a more delicate oolong or Earl Grey tea, one survey showed that 68 per cent of people in the UK drink two or more cups a day. Tea is an essential part of many people's daily life, helping them to wake-up, relax and unwind.

It is estimated that tea was first used as a medicinal drink in China in 2737 BC. China is the largest producer of tea followed by India and Kenya. But it's Ireland that drinks the most tea per capita, followed by the UK.

THE IMPACTS

Tea bags are typically made from paper and sealed with polypropylene, a plastic – a fact that explains why gardeners reported very slow decomposition of tea bags in their compost heaps. A 2010 *Which?* gardening study in the UK revealed tea bags from top manufacturers were only 70–80 per cent biodegradable. In some tea bags, plastic is contained in the mesh of the bag itself as well as in the glue that holds it together. Given the increase in public awareness about plastic, a number of tea producers have set targets for removing the plastic from their tea bags.

The environmental footprint of a tea bag isn't just down to the plastic – it's also the resources that go into growing, processing and transporting tea. A 2012 study revealed that over 300 litres of water are required to grow enough tea for 25 tea bags and when grown in water-scarce areas, tea places a stress on groundwater supplies.

A 2017 study published in the Oxford Research Encyclopedias states that the intensive, monoculture production of tea has a high impact on the environment. It typically

THE FIRST TEA BAG

The tea bag was invented by accident in America in 1908 by Thomas Sullivan, a New York tea merchant. He sent samples of tea to his customers in small silk bags; they assumed the bags were to be used in the same way as the metal infusers to make tea. Sullivan then refined his approach to use gauze rather than silk and began commercial production in the 1920s. However, they didn't become popular until the 1950s due to wartime materials shortages. Once they hit the shelves, tea bags were a winner with tea drinkers due to their convenience. In the early 1960s, tea bags made up less than 3 per cent of the British market rising to 96 per cent of the market by 2018.

requires the use of pesticides and inorganic fertilisers that create environmental hazards, cause water pollution and threaten biodiversity. The repeated application of fertilisers and herbicides also contributes to soil degradation, and the acquisition of new land for tea growing contributes to deforestation and habitat loss. The processing (drying and fermenting) and transport of tea relies on the burning of fossil fuels, which contributes to climate change. So, a cup of tea with zero impact is not possible – but it is possible to make choices that are better for you and the environment.

WHAT YOU CAN DO

- Use loose tea leaves and make a pot of tea – many people say this makes the best cup of tea because brewing tea in its loose-leaf form allows the hot water to infuse the whole leaf, producing the freshest, fullest flavour possible. Buy and use a metal tea infuser if you just want to make one cup of tea at a time.

- Buy plastic-free tea bags in minimal packaging. Ideally choose certified compostable tea bags and put them with your food waste in a brown (organic) bin. Don't expect your home composter to be able to cope with tea bags - they break down more slowly than food waste.
- Avoid overpackaged and individually wrapped tea bags.
- Look out for Rainforest Alliance or fairtrade certified tea to ensure that the tea has been produced to high environmental standards – meaning less impact on the environment.
- Uphold standards by choosing fairtrade or ethical tea partnership certified tea – such products provide better working conditions and pay for people working all along the supply chain from the growers and pickers to the factory workers.
- Choose certified organic tea that is grown and produced without synthetic fertilisers, pesticides or herbicides, so that it is better for biodiversity and protects the health of the growers and pickers.

THE FIRST PLASTIC-FREE TEA BAG

One company working hard to make sure all their tea bags are plastic free, as well as ethically sourced, is teapigs. Nick Kilby and Louise Cheadle set up teapigs to bring more varieties of whole-leaf tea to tea lovers. Their quest for flavour and quality also led them to care about the environmental footprint of their tea and the lives and wellbeing of the farmers that grow it.

From the start, the company put an emphasis on whole-leaf tea – so no standard tea bags. Instead they use pyramid-shaped tea bags that they call 'tea temples' to hold their tea. These temples are made from corn starch, which is biodegradable, as is the string, and the label is made from paper printed with vegetable oils. The temples are heat sealed, so there's no plastic glue either.

The company is transparent about what biodegradable means for their tea and don't recommend putting them in your household compost as they will take too long to break down, but they can go in with your food waste for industrial composting.

Packaging-wise, their cardboard cartons are made from FSC-certified cardboard, which is recyclable, and the clear, inner bags are made of a material called Natureflex. It might look like a plastic bag but this cellulose product is compostable in a home or an industrial composter. The company also makes loose tea in aluminium cans, which can be recycled or repurposed.

Best of all, teapigs provides clear, honest information to their customers about their impact on the environment and how they work with their tea-growing communities.

In 2018, teapigs were the first tea brand to be awarded the Plastic Free Trust Mark – certifying that the product and the packaging are plastic free. The certification is awarded by A Plastic Planet, a not-for-profit organisation based in London.

THE REAL PRICE OF COFFEE

I love coffee and drink it every day. I am part of a massive market for a commodity that is traded so frequently that the price changes every three minutes. Yet, almost 61 per cent of coffee growers are selling their coffee at a loss.

According to Fairtrade (see more below), approximately 80 per cent of global sales of coffee are now attributed to just three multinational corporations; and innovations such as coffee pods and capsules have created bigger profit margins for big brands with little benefit to the coffee farmers.

THE NEED FOR FAIRTRADE

Fairtrade is the world's largest and most recognised fairtrade movement with branches in different countries. It works with businesses, consumers and campaigners to get a better deal for farmers and workers, and promotes sustainable livelihoods and production systems. When farmers are squeezed because they receive so little money when selling their crops, they are forced to cut corners; it's all too easy to employ children, pay themselves and their workers less and reduce measures to protect the environment. This creates a vicious cycle of increasing poverty, degraded soil and, for us, poorer-quality coffee.

A study carried out by the organisation True Price and Fairtrade International in 2017 found that 100 per cent of Kenyan farmers, 25 per cent of Indian farmers and about 35–50 per cent of both Indonesian and Vietnamese farmers do not earn enough to live on from growing and selling coffee.

MAKING FAIRTRADE WORK

About 125 million people are dependent on coffee farming around the world. If the coffee farmers were paid a fair price for their product, the impact on the wellbeing of their communities would be transformative – and just from paying a fair price for a product we want and value.

You can help achieve that goal by buying fairtrade coffee (and tea, bananas, sugar and cocoa) so that farmers receive a fair price for their produce and, on top of that, a premium for their product that they can invest in agricultural tools and supplies, training, health and education or environmental protection.

The Rainforest Alliance is another organisation that certifies products, including tea, coffee, cocoa, palm oil and cut flowers as environmentally, socially, and economically sustainable, using an easy to spot symbol with a tree frog on certified products. They are now working on an upgraded certification programme as a result of their merger with UTZ (a certification programme for tea, coffee and cocoa), so look out for new information and a revamped logo on your bag of coffee.

Whether you prefer a flat white, Americano or espresso, in the UK we drink about 95 million cups of coffee per day.

COFFEE MAKER

Many of us rely on a good cup of coffee to get the day started. And the way in which we choose to make our coffee and the machines we use can affect the environmental impact of our cup of joe.

People have been brewing coffee for centuries (the earliest reports are of the Turks brewing coffee in 575AD) and some old methods live on to this day. In Ethiopia a traditional earthenware coffee pot heated over a charcoal stove is still used in coffee ceremonies.

The first coffee percolator (a metal stovetop design) was patented in the US by James Nason in 1865, and the first drip coffee machine, made using blotting paper, was invented by an Italian woman – Melitta Bentz – in 1908. The cafetière or French press didn't come along until 1929.

Nowadays, there are international coffee conventions and schools for baristas. We have an array of choices, including the most recent invention – the coffee machine that uses coffee pods or capsules.

THE IMPACTS

All electric coffee makers consume energy, as does the alternative of boiling the kettle to make an instant coffee or to fill a French press. So, in terms of their environmental impact, how do these methods compare?

If the coffee machine has a standby power mode it will be consuming energy, even when inactive. This so-called 'vampire energy' use will add to your electricity bill (see also page 89). Espresso machines use a little more energy per cup of coffee than drip coffee machines. Push-button coffee machines that make you a single cup of coffee may have a reservoir that is keeping water warm – and using energy to do so.

From a materials points of view, electric coffee machines are made of metal and plastic

and are electronic waste that has to be WEEE recycled at the end of their life. Many of the parts are valuable and can be separated for reuse and recycling if you take the appliance to the recycling centre or to retailers who accept old appliances for recycling.

The additional waste associated with coffee making and coffee machines comes down to how the coffee enters the machine. Most drip coffee makers use single-use paper filters and these can be composted along with the coffee grinds. Coffee pods and capsules, however, create more waste than drip or espresso coffee machines as each serving of coffee is contained in a plastic or aluminium capsule. The production of capsules consumes natural resources and requires energy – so their environmental footprint is larger than a bag of coffee beans or ground coffee.

Generic coffee pods are not recyclable in your household recycling as they contain coffee grinds, aren't easy to clean and are usually a mix of plastic and aluminium. Brands such as Nespresso offer a specialised collection and recycling service for their pods.

Avoid coffee pods claiming to be biodegradable in preference for those which are compostable (biodegradable ones will take a long time to decompose in a landfill, releasing climate-change-causing methane gas in the process). Make sure they are labelled compostable and ensure they go for industrial (kerbside) rather than home composting.

WHAT YOU CAN DO

- Compost your paper coffee filters and coffee grinds. Grinds make an instant compost – you can add them directly to your plant pots and garden as a soil conditioner. Plastic, cloth, or plastic-coated coffee filters that are at the end of their life must go into the rubbish. So, look out for reusable coffee filters made of natural fibres.
- Read the manual or watch a YouTube video about your coffee machine and learn how to use it in 'eco' mode to reduce energy use. Setting the temperature and reducing the standby time can save energy, as can turning the machine off at the plug when not in use.
- Choose a coffee machine if you're buying a new one that uses whole beans or ground coffee and not pods. If you have a pod machine, choose reusable coffee pods. You can also look online for ideas on emptying, refilling and reusing your old pods.
- Maintain your coffee maker and descale it regularly so that it heats efficiently and so uses less energy.
- Keep making coffee at home – having a coffee at home is more environmentally friendly than driving to a café or buying coffee in a disposable cup, even if that cup is compostable!

SUSTAINABLE FOOD

It often feels like buying food has never been more complicated! You want to buy organic vegetables but they are wrapped in plastic. You don't know which is worse for the planet, a steak or an avocado? It can seem like an impossible challenge to avoid sugar, salt, GMOs and palm oil and to buy local, seasonal and healthy food – and on top of that, all at an affordable price. Caring about animal welfare, chemicals in food, fairtrade, nutrition and the carbon footprint of food makes food shopping a minefield.

We live in a world where, according to the United Nations, over 800 million people are hungry and undernourished, yet one-third of all food produced is wasted and goes uneaten. The value of food waste is a staggering US$940 billion per year. To put that in perspective, in 2018 the official aid given by the thirty members of the OECD to developing countries (for health and education, food aid and humanitarian assistance) was a mere US$153 billion. So, the cost of food waste dwarfs commitments to global solidarity and human development.

FEEDING THE WORLD

Agriculture and the production of food is a huge contributor to climate change – it's responsible for up to one-quarter of global greenhouse gas emissions; ironically, this sector is also highly vulnerable to the impacts of climate change. Estimates reveal that by 2050 there will be 9 billion people to feed on a planet that is warmer than today with more extreme weather events (droughts, floods and storms) all interfering with food production and supply chains. Furthermore, changing seasonality, declining soil fertility and water availability will reduce the yields of crops such as rice, maize and wheat just at the time when there are more mouths to feed.

THE UBIQUITOUS INGREDIENT – PALM OIL

Palm oil is in everything from pizza, chocolate and doughnuts to soap, shampoo and lipstick. It comes from the fruit or kernel of the palm oil tree, which is native to Africa but is now grown predominantly in Malaysia and Indonesia, where native rainforest has been cut down to make way for palm oil plantations. This deforestation contributes to climate change and to biodiversity loss, as orang-utans and pygmy rhinos lose their homes. Silence reigns supreme in a palm oil plantation because there is little wildlife apart from rats and snakes whereas in a nearby rainforest, the noise of insects, birds, gibbons and other animals is deafening.

Palm oil can be produced more sustainably, though, and provide decent jobs if it is produced differently, supporting smallholders and enabling them to manage the landscape to conserve intact rainforest. The Roundtable on Sustainable Palm Oil (RSPO) is a global organisation working to reform the industry. As far back as 2005, it implemented its pilot scheme. Having learned lessons across several years, it designed a set of principles and criteria for smallholders to comply with in order to receive RSPO accreditation, the latest version of which was updated in 2018. Look for the RSPO Certified Sustainable Palm Oil mark on products.

EAT MORE SUSTAINABLY

So, what is the solution? Some changes will need to be happen at the policy and international level to reduce carbon emissions and regulate the production and trading of food, but there is a lot that we, as individuals, can do, too. Here are some of the rules I use when food shopping:

- Buy as little processed food as you can – this will help you to steer clear of food additives, palm oil and excess sugar, salt and packaging.
- Buy food as locally as you can. If you have a local greengrocer, butcher, fishmonger or farm shop (or box delivery) buy from them when you can. If you mostly shop in a supermarket, you may have less choice, but you can find local foods and encourage them to stock more.
- A little of everything and not too much of anything. So, eating meat every day is too much, especially if this means your chicken, pork and beef has been mass produced on industrial farms. Instead, buy the best you can (free range, local, organic) and eat it once or twice a week rather than every day.
- Eat fruit and vegetables in season. We don't need to be able to eat raspberries and mange tout all year. Buying food in season reduces the carbon footprint associated with its transport.
- Ask where your food comes from. The traceability of food is improving all the time, so don't be afraid to ask.
- Learn how to cook. You will be healthier, save money and be kinder to the planet. Cooking for yourself means you can also choose to use cheaper ingredients to make your money go further.
- Learn to love leftovers! Leftovers are the makings of tomorrow's lunch or dinner. Freeze leftover juice and fruit for smoothies year round.
- Shop smart to avoid waste. Make a list and stick to it. Don't buy multipack offers unless you know you will use it all. Buy a big tub of yoghurt and decant it into reusable containers to take to school or work.

FOOD CARBON FOOTPRINTS

The figures here show the comparative carbon footprints (expressed as gCO_2e) to relate the environmental impacts of each type of food. Local and seasonal is key in reducing the carbon footprint of most foods. Meat has a higher carbon footprint than vegetables and local produce is always best.

On a tour of a supermarket, we find in the greengrocery section, a local apple clocks up **10** but the average is **80**. A punnet of strawberries in season is just **150** but out of season and air-freighted shoots up to **1,800**! Similarly, a bunch of seasonal asparagus is **125** but this sky-rockets to **3,500** out of season and flown into the UK. Tomatoes show a similar story – 1kg of organic, local and seasonal tomatoes is **400**, compared with organic, off season and grown in the UK of **50,000**. The household staple of a pint of milk is **723** and a loaf of bread **800**. When looking at the fish counter, 1kg of mackerel is **500** and 1kg of prawns is **10,000**. A quick look at the burger offer reveals that a veggie burger is **1,000** and a regular cheese burger is **2,500**.

Over 100,000 metric tonnes of aluminium packaging sold in the UK was recycled in 2018 and 95 per cent of aluminium packaging collected in the UK is recycled within Europe.

ALUMINIUM FOIL AND TRAY

Aluminium foil is used to package everything from chewing gum and butter to takeaway food. We also cook with it – from roasting the Christmas turkey to baking fish. Aluminium is 100 per cent recyclable. In fact, recycling aluminium saves more than 90 per cent of the energy costs required in its production.

Aluminium foil experienced a period of rapid growth between the 1950s and the 1960s when TV dinners packed in compartmental trays began to reshape the food products market. Today, aluminium features not just in foil and trays but in drink cans, sachets, pouches, lids, wrappers, blister and strip packs.

Aluminium is lightweight and durable. It was first used in 1910 by Robert Victor Neher in Switzerland. By 1911, Toblerone chocolates were being wrapped in aluminium foil, followed in 1912 by the food company Maggi who used it to encase stock cubes.

Aluminium – the third most abundant element in the Earth's crust – is primarily extracted from bauxite, with most aluminium coming from China (which accounts for half the world's production), followed by Russia, Canada and India. China is also the main consumer of aluminium, accounting for over 40 per cent of global sales.

The majority of aluminium packaging is used for food and beverages (75 per cent) with 7 per cent used for pharmaceuticals and 8 per cent for cosmetics, among other things. Because aluminium has good barrier properties, it forms a complete barrier to light, oxygen, moisture and bacteria, it is the preferred packaging for coffee, tea, spices and other aromatic goods and it plays an important role in food preservation and preparation hygiene. It is also used to make 'disposable' cooking and baking trays.

THE IMPACTS

The production of aluminium foil and trays requires the extraction of natural resources from the Earth which uses huge amounts of energy (nine times as much energy as is required to make steel) and produces by-products such as 'red mud' sludge, which is stored in reservoirs for treatment. In Hungary in 2010 a dam containing a red-mud reservoir collapsed and released a 1–2-metre wave of toxic waste over the village of Kolontár and the town of Devecser, killing ten people and injuring 120.

There have been some concerns expressed about a possible link between Alzheimer's disease and metals including aluminium but research is ongoing.

WHAT YOU CAN DO

- Find alternative food covers. Try a beeswax wrap instead (see page 27) or put a plate on top of a bowl of food or a bowl over food on a plate. You can also buy reusable silicone covers for plates and bowls.
- Go wrapping free and use a lunch box or substitute foil for compostable greaseproof paper (see page 28) or recycled paper towel. If you do use foil, be sure to reuse it several times, then wipe it clean and recycle it.
- When cooking, use a real tin or baking sheet and reduce waste. You will have to wash it but that is no real hardship, is it? Aluminium trays (such as those for a Christmas turkey) have to be cleaned after use anyway in order to be recyclable. If you are covering food in the oven use a baking sheet or a pan with an ovenproof lid instead.
- Clean and recycle. Only recycle foil that is spotlessly clean (butter wrappers and the foil used to roast meat are pretty impossible to get clean and should go in the bin). Scrunch clean foil into a ball and recycle alongside aluminium cans. If the foil is lined with paper or plastic it cannot be recycled.
- Purchase carefully. Crisp packets cannot be recycled as the foil is layered with plastic or paper. Likewise tubes of crisps that have an aluminium base; their mix of materials means they can't go in household recycling. Look out for specialised recycling services for crisp packets and tubes from TerraCycle.

Additional facts

The European Aluminium Foil Association reported that 942,500 tonnes of aluminium foil were produced in Europe alone in 2018.

Every minute, an average of 113,000 aluminium cans are recycled in the US alone.

Aluminium cans are the most valuable commodity to recycling companies: in 2018 a tonne of aluminium cans was worth £1400 compared with £188 per tonne for PET plastic.

More than 1.2 billion metres (745,000 miles) of cling film is used in households across Britain every year – enough to go around the circumference of the world thirty times over.

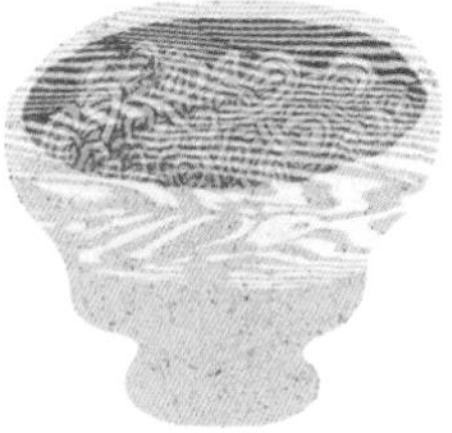

CLING FILM

Cling film has been revolutionary – it works well, is low cost and is convenient. But who hasn't experienced the frustration of trying to find the start of it on the roll and then struggled to untangle it? It's so good at its job that it has a tendency to stick to itself as you unwrap it off the roll – but perhaps there are convenient alternatives that are also good for people and the planet.

Cling film covers bowls and plates of food in homes, hotels and restaurants around the world. It plays an important role in food hygiene and preservation, acting as an oxygen barrier to prevent food spoiling, plus it keeps flavours and smells from cross-contaminating other foods nearby. It also has uses outside the kitchen – to wrap wounds and burns, to attach cold packs to athletes with sports injuries and an industrial-strength cling film to wrap luggage for air travel.

THE IMPACTS

Cling film is a soft plastic and is not recyclable in most parts of the world. It has to go in the bin for landfill or incineration (see also pages 46–49). It takes hundreds of years to decompose and out in the environment it breaks down into microplastics that can be eaten by wildlife and enter the food chain.

Chemicals called phthalates are added to cling film to make the plastic soft and elastic and to enable it to cling to surfaces. Some researchers have linked phthalates to health risks including asthma and obesity. A study conducted in the UK in 2012 found that 75 per cent of bread products contained one or more phthalate,

FROM PLANE TO PLATE

Cling film was originally made from polyvinylidene chloride (or PVDC for short), which was accidentally discovered at Dow Chemical Company in 1933. It was first used as a spray to protect plane and car upholstery, before being used to make insoles for jungle boots. In 1949, the thin, clingy plastic wrap sold as Saran Wrap in the US was put on the market for wrapping food. Nowadays, though, cling film is made from polyethylene, which is cheaper and easier to produce than PVDC and also safer for people.

thought to have come from the packaging. Since these levels of phthalates were low, regulations have not yet limited their use in packaging in the European Union.

Cling film is made from nurdles which are tiny plastic pellets, smaller than a lentil. When nurdles escape from factories or shipping containers they end up in rivers, on beaches and in the sea. A beach clean and nurdle hunt on a beach in Scotland collected over 540,000 pellets in one day in 2017.

To be safe and to ensure there is no transfer of chemicals from cling film into your food, it is best not to have cling film in direct contact with food when you microwave it.

WHAT YOU CAN DO

- Place a plate on a bowl or a bowl on a plate for leftover food at home – there is no need for cling film. Plus, it saves money and time.
- Store leftover food in any reusable container (plastic, glass or metal). These will stack better than containers covered in cling film, maximising use of space in the fridge.
- Repurpose empty glass jars – glass does not affect the taste of food so it is a good alternative to plastic. Reuse old ice-cream containers and other reusable packaging to store and freeze food.
- Opt for greaseproof paper (see page 28) – it's great for wrapping sandwiches and cakes. Choose unbleached, compostable, greaseproof paper rather than the more common coated baking parchment.
- Use a lunch box with compartments to separate sandwiches from biscuits and fruit in a packed lunch.
- Buy some beeswax wraps and use these instead of cling film. Such wraps are made of a canvas covered in beeswax and are sticky enough to form a seal when heated with the warmth of your hand. They can be washed and dried and used over and over again.
- Make your own beeswax wraps out of pieces of cotton (old shirts and blouses) and melted beeswax. Tailor the size of the wraps to suit different sized plates and bowls.
- Unfortunately beeswax wraps can't take the place of cling film in commercial environments, such as hotels, restaurants and canteens, for health and safety reasons. Compostable plant and wood-based film is in development but it is not yet commercially available.

Although parchment paper is designed to be single use and disposable for convenience, it can be wiped and reused several times.

BAKING PAPER

If you like baking then you will, no doubt, use a baking paper to line tins when making cakes and cookies. Whether you choose parchment paper, greaseproof paper or non-stick baking paper, these papers are designed to be used once and then thrown away.

Previous generations of bakers only had access to greaseproof paper, which is a paper made by beating paper fibres until they become more firm and tightly bonded together. The resulting paper is dense and less absorbent, meaning that it resists oil and grease. So, it's great for wrapping butter or greasy foods and for cooking. Its downside is that you still need to add butter or oil to ensure that baked goods don't stick.

A new cousin to greaseproof is non-stick baking paper or baking parchment. It tends to be more expensive but is essentially greaseproof paper with a non-stick coating. Such parchment paper can be used at high temperatures and, perhaps unknown to many cooks and bakers, be reused several times. It is super-convenient and is available to buy in rolls or even pre-cut to fit typical baking sheet and tin sizes. The non-stick coating is made from silicone (a kind of synthetic rubber). Silicone should not be confused with silicon, which is a naturally occurring mineral.

THE IMPACTS

Since baking papers are made from paper, and as demand for baking paper increases as incomes rise around the world, more trees are needed to produce the paper required. Over 50 per cent of the world's industrial logging is used to make paper, making paper a contributor to deforestation where sustainable forest management is not assured.

Furthermore, paper manufacturing is the largest user of water per kilogram of finished product, and the paper industry is one of the leading emitters of greenhouse gases in the manufacturing sector due to its consumption of energy.

Baking paper can be made from either bleached or unbleached paper (see page 194). Given concerns about the environmental impact of chemicals in the wastewater from paper manufacturing, unbleached paper is now being used to make many baking papers.

Since greaseproof paper is just beaten-up paper, it can be composted after use. Parchment paper, however, is coated with silicone so its suitability for composting depends on the type of silicone used.

Some silicones are based on a silicon and oxygen compound, making them organic rather than plastic. Parchment papers that use this organic type of silicone are labelled compostable and have been tested to comply with the standards for compostability. That means that the paper breaks into small enough fragments over a 24-week period to meet industrial composting standards (EN 13432).

PFAS (poly- and perfluorinated alkyl substances) are used in some baking and greaseproof papers – such as microwave popcorn bags – because they repel water and fat. Certain PFAS are known to persist in the environment and to accumulate in humans. As a result, some companies have phased out their use voluntarily, but these substances are still produced and used in some products. There are fears, though, that the alternatives being used to replace them are under-regulated and may also be hazardous to the environment and human health.

WHAT YOU CAN DO

- Consider a zero-waste option and simply grease your baking tin really well – just as your granny used to do.
- Look for unbleached greaseproof paper – you will need to grease it lightly with oil or butter if baking – but it can be composted, which keeps it out of the waste cycle and prevents it from going to landfill where it will produce methane gas as it slowly decomposes.
- Choose compostable baking parchment made from unbleached FSC-certified or recycled paper. Wipe and reuse it several times and then put it in your brown bin for industrial composting after use.
- Have a closer look the next time you buy baking paper. Are you getting greaseproof (which is best) or baking parchment (with its extra coating)? Does the package tell you anything about the non-stick coating used?
- Transform old greaseproof paper into firelighters.
- Try out a reusable silicone baking tray. These non-stick baking items can be used thousands of times and mean you don't need to use parchment paper at all.

These bags have so many uses and are so convenient that the average family in the US uses 500 of them each year.

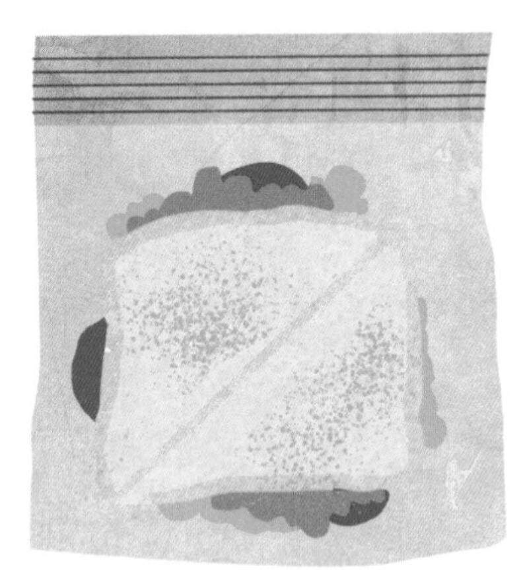

PLASTIC FOOD BAG

If we look at children taking packed lunches to school, even if we estimated that only half of them take sandwiches for the roughly 200 lunchtimes in a school year, we're looking at over 800 million disposable food wrappings in a year. Now, if we turn our attention to the world of work, the numbers become even more massive.

Back in the day you put your sandwich in a lunchbox, wrapped it in some paper or put it in a paper bag to take it to school, work or on a picnic. But then came the invention of plastic and the now familiar plastic food bags.

Today these bags come in every conceivable size – for storing chopped fruit and veg, sandwiches, leftovers and food for freezing. In fact, they are so versatile people now use them to hold all sorts of things other than food – cosmetics and medication when travelling, to organise a suitcase, to hold craft materials and crayons for kids, to marinade food – the list goes on. According to *Vogue* magazine, there is 'no end of uses for those great Ziploc bags'.

THE IMPACTS

Plastic food storage bags of all types are intended as single-use, disposable items (see also pages 134–138 and 46–49). They are made of plastic, primarily low-density polyethylene (LDPE) and linear low-density polyethylene (LLDPE).

Like all plastics, LDPE and LLDPE are very durable and will take hundreds of years to decompose. Soft plastics, such as plastic bags, are not accepted in household recycling bins. Some supermarkets and recycling centres have specific recycling points for plastic bags and if your plastic food bags are clean and dry you can use these facilities, too.

WHAT YOU CAN DO

- Use a lunchbox for packed lunches and sandwiches. Lunchboxes with compartments allow you to separate your fruit from your bread roll, and they can be used over and over again.
- Choose stainless-steel and glass storage jars as good alternatives to plastic bags for storing food in the fridge, and stainless-steel containers also work well in the freezer.
- Look for reusable and washable cloth sandwich bags.
- Go for a paper bag if you need a disposable option. These are easy to source, in a shop or online. Compostable sandwich bags must be put in a brown bin for industrial composting. In a regular rubbish bin they just become regular rubbish destined for landfill.
- Reuse any plastic food bags, in particular resealable ones that tend to be a thicker plastic. Wash them out, dry them (just leave them hanging over a large kitchen utensil overnight) and use them again and again.
- Reuse all possible packaging. Some foods (such as muesli) come in durable resealable packaging. Just rinse it out and use it again. You can even reuse the bags that bread and bagels come in to store food in the fridge or freezer or to wrap a sandwich for lunch.

THE ORIGINS OF ZIPLOC

The first plastic food bags on a roll appeared in the late 1950s as a convenient, disposable and leak-free alternative to paper bags and became commonplace by the 1960s. The 1960s saw the arrival of resealable plastic bags with a Ziploc for the purpose of boiling food – put your veggies inside and submerge the bag in a pan of boiling water to cook. By the 1970s Ziploc bags, created by Steven Ausnit, were sold for storing food. The first Ziploc was a press and seal mechanism, later superseded by the double zip to improve the seal and then the sliding zip to make opening and closing even easier.

Research has shown that kitchen sponges can have more germs than your toilet and are comparable in the density of bacteria they host to the human intestinal tract!

WASHING-UP BRUSH AND SPONGE

While survival specialists may be able to clean pots and pans with sand and water, most of us stand at a sink and use a sponge, a brush or a cloth to wash the dishes. Sea sponges have been used for cleaning purposes since the Middle Ages in Europe, but modern dish sponges have a very different connection to the sea – they often end up there as waste and not as part of the biodiversity.

Nowadays washing-up sponges are made from polyurethane foam, a material invented at the start of World War II and initially used as a rubber alternative to protect and cushion wood and metal. In the late 1950s, flexible polyurethane foam was developed to expand its number of uses, from insulation and cushioning in furniture and cars to cleaning sponges.

THE IMPACTS

Polyurethane foam has received much attention in the past due to its detrimental association with the ozone layer. CFCs or chlorofluorocarbons were once used as a blowing agent to make the foam until they were found to be contributing to the hole in the ozone layer and the Montreal Protocol in 1987 restricted their production and use. Overall the Montreal Protocol is a success in international environmental law as it has worked. But it replaced CFCs with HFCs (hydrofluorocarbons). HFCs don't harm the ozone layer as they do not contain chlorine. But since they are greenhouse gases, HFCs are being phased out. That said, an investigation in 2018 by the Environmental Investigation Agency (in the US) uncovered the widespread and illegal use of CFC-11 in the polyurethane foam insulation sector in China since 2012, despite its production and use being banned.

Plastic sponges and washing-up brushes have also been in the spotlight as they cause plastic pollution. Neither is recyclable and both hang around in the environment for a very long time. Plastic foam, of the kind found in washing-up sponges, is in the top ten most commonly found pieces of litter on European beaches.

So, given all of the above, why are we still so wedded to plastic dishwashing sponges? It would be easy to imagine that it is because they provide a superior clean. But no, the facts around the germs lingering in a washing-up sponge are shocking. A study published in 2017 found that used kitchen sponges are home to 362 different species of bacteria in the sponges examined, and the density of bacteria reached up to 45 billion per square centimetre. Such reservoirs of bacteria can be spread across your kitchen as you use them and can lead to cross-contamination of hands and food, which is a cause of food-borne disease outbreaks.

WHAT YOU CAN DO

- Use a dishwasher. Yes – it is more sustainable and doesn't require a dish-washing sponge! A life cycle analysis from 2013 compared handwashing with using a dishwasher and found that washing dishes by hand uses more water. Depending on the size of your household, manual dishwashing can use an average of up to 50 litres of water, 2.60kWh of power and about 60 minutes to complete per wash. On the other hand, washing with a new, efficient dishwasher uses only 6.5 litres of water and 0.67 kWh of energy, plus you can load and unload the dishes in less than 15 minutes – so it's win-win!
- Stop buying plastic sponges and use a sustainable alternative such as a loofah or cotton cloth that you can wash and reuse without the hygiene risks of a sponge. Avoid using cloths made from synthetic materials, such as polyester, as they release microfibres when washed. Loofahs are made from a dried-out vegetable that looks like a cucumber. I use them and think they do a great job as both a sponge and a scourer.
- Swap your plastic washing-up brush for a wooden one. Some have replaceable heads and the wooden part can be composted once you cut off the bristles. Brushes are also more hygienic than a sponge, as they tend to stay drier when not in use and don't have as many nooks and crannies to harbour bacteria.
- Invest in non-plastic scourers. Try a wooden brush with hard bristles, a stainless steel or copper metal scourer (both are recyclable) or for a slightly gentler approach try a loofah. In Japan people use a *kamenoko tawashi* made from coconut fibres to clean fruit and veg as well as pots and pans.
- Source sponges made from recycled plastic. These have the advantage of not being made from virgin plastic – but disposal is still problematic as they are not recyclable.
- Repurpose nasty old sponges for other dirtier jobs. Get as much use as possible from them before they go in the rubbish by using them to clean bins and boots.

Additional facts

Interestingly, a study in 2001 found that using antibacterial washing-up liquid had no effect on reducing the pathogens in dishwashing sponges.

25 per cent of the phosphorus in wastewater comes from detergents, including washing-up liquid.

WASHING-UP LIQUID

In 2014, Europeans did an average of 4.3 dishwasher loads of washing up per week. In the UK and Ireland 58 per cent of people own a dishwasher and typically run 5.2 loads per week – that is almost a load for every day of the year – and about 270 dishwasher tablets.

Washing-up liquid is, of course, closely related to soap (see pages 124–125), but falls into a category of products called detergents. During World War I and World War II there was a shortage of the animal and vegetable fats and oils that were used to make soap at that time. Chemists had to innovate with other raw materials that were 'synthesised' into chemicals with similar properties – known as detergents.

Detergent molecules are hydrocarbon chains (made from crude oil) with one end that's attracted to water and another end attracted to oil. The oil-grabbing end attaches itself to food on the plate and surrounds it with the water-grabbing ends, so that the engulfed particles become water-soluble and can be rinsed away.

THE IMPACTS

When water runs off the land, it can contain large amounts of nutrients, such as phosphorus and nitrogen, from detergents or fertilisers. When these chemicals end up in nearby ponds, lakes or rivers, they enrich that water (often associated with bright-green algal blooms and a stark reduction in wildlife) – it's a phenomenon known as eutrophication.

A study published in 2015 looked at the contribution that household chemicals make to water pollution. It revealed that phosphorus from detergents (where it's in a form known as phosphate), including washing-up liquid, contributes to eutrophication from household sewage.

As a result, the EU banned the use of phosphates in detergents in 2017, and now many companies are following the example set years ago by brands such as Ecover and are reducing or removing the phosphates from their detergents. International companies such as Unilever reported that by the end of 2018 it had eliminated phosphates in 100 per cent of its dishwashing products (and reduced by more than 95 per cent the use of phosphates in their laundry powders globally).

Another environmental impact is the fact that most washing-up liquid comes in a single-use plastic bottle, usually made from virgin plastic. Plastic bottles can cause pollution and enter the food chain as they break down into nano- and microplastics (see also pages 134–135). The plastic is recyclable but in reality that bottle could be used over and over again – it does not need to have just a single use.

Ecover washing-up liquid comes in 100 per cent recycled plastic bottles and is currently (at time of writing) the only big brand that offers a refill service.

In 2017 Fairy Liquid launched the Fairy Oceans Plastic Bottle, developed in partnership with global recycling firm TerraCycle, which are made from 100 per cent recycled plastic, including 10 per cent ocean plastic (plastic waste recovered from the ocean and recycled), and are fully recyclable. However, the bottle is not yet used across the entire range of washing-up liquids.

WHAT YOU CAN DO

- Buy environmentally friendly washing-up liquid from a well-established eco brand. Look out for the following attributes of a truly eco-friendly washing-up liquid: plant-based (rather than petroleum-based) surfactants; minimal toxicity to aquatic organisms (OECD test 201 and 202); biodegradable within 28 days and completely (OECD test 301F); clear information on how to refill or recycle the packaging; and a list of all ingredients in language you can understand.
- Refill your bottle. Look for refill shops, a stall in your local market or a home delivery service that offers refills of eco-friendly washing-up liquid. In this way, you can use the same washing-up liquid bottle for years.
- Try some solid, all-natural dishwashing soap in block form and cut out bottles altogether. You will need to look in a specialist store or online shop to find this.
- Buy in bulk. If you can't find a local refill solution then buy the biggest container you can and decant as necessary to reduce the volume of packaging.
- Recycle all empty bottles you are not reusing. Make sure they are clean and dry and put them in your household recycling bin.

Households worldwide get through 6.5 million tonnes of kitchen roll every year. That is a lot of trees!

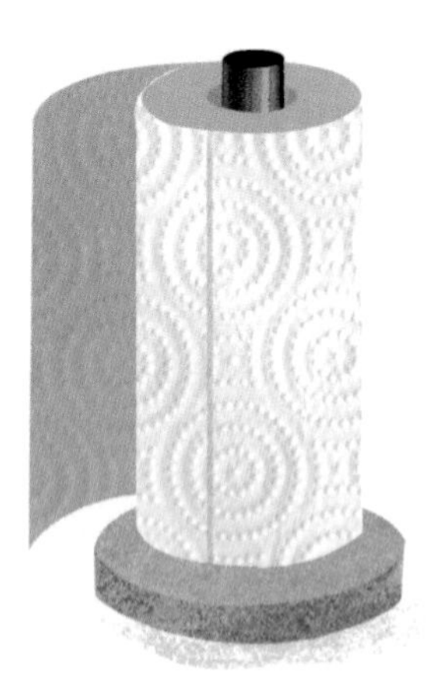

KITCHEN ROLL AND PAPER NAPKINS

People used to wipe up spills with a cloth, wring it out and rinse it, and wipe again. Not a huge hardship, but nevertheless convenience and hygiene drive the global sales of kitchen roll. In the case of paper napkins, it is really all about convenience and time saving; cloth napkins do a better job, and probably one will do where a handful of paper napkins would be needed, but in busy homes and restaurants it is more convenient to go for the disposable option.

Tissue papers as they're known technically, which comprise everything from tissues and kitchen roll to paper towels, were first manufactured by the Scott Paper Company in the 1880s. Originally for medical use, they were later repurposed for drying hands of the students at a school in Philadelphia to reduce the spread of colds. In 1931 the company started to market paper towels for kitchen use in the US; and now the US uses almost half the world's supplies of kitchen towel.

Demand for kitchen roll has grown year on year to become a multi-billion dollar industry. In 2017, the average American spent three times as much money on kitchen roll as the next biggest consumer, Norway. Kitchen roll, paper towels for drying hands and paper napkins are all manufactured to have different properties from absorption to durability and softness and that helps the market to grow.

THE IMPACTS

Kitchen roll and disposable napkins are made from paper – so all the environmental impacts associated with paper production apply (see pages 193–195). From deforestation and water consumption to the pollution associated with pulping and bleaching, the production of 'paper towels' impacts both the environment and human health.

From a waste perspective, disposable kitchen roll and paper napkins unavoidably create waste. They are designed to be used once and thrown away. A report by the US's Environmental Protection Agency in 2015 found that the US generated about 3.3 billion kg of waste per year consisting of paper towels and other 'tissue' materials, such as toilet paper. Meanwhile in the UK 34 per cent of people incorrectly think dirty kitchen roll is recyclable and put it in their recycling bin where it contaminates other materials for recycling.

WHAT YOU CAN DO

- Use a cotton cloth for mopping up all spills, rinse and go again. Simply wash as you need to maintain hygiene standards.
- Opt for a reusable lunch box then you won't need to wrap sandwiches or cake in kitchen roll for packed lunches.
- Wipe kids' faces and hands after dinner with a wet face cloth or flannel.
- Buy or make your own reusable kitchen towels. Find these for sale in package-free shops; they are good quality and can be used over and over. Swedish dishcloths made from cotton and FSC-certified cellulose are a good option, too, as they absorb twenty times their weight of liquid and they claim that each cloth replaces seventeen rolls of kitchen paper and lasts over nine months and 200 washes.
- Make your own reusable kitchen towels from old towels, face cloths or soft sheets.
- Switch to cotton napkins. Buy organic cotton ones, make your own or discover pretty pre-loved ones from a vintage shop. They are a whole lot nicer than paper versions and are better for the planet.
- Make sure any kitchen roll you buy is made from 100 per cent recycled paper or is FSC-certified. And always recycle the inner cardboard tube.
- Replace kitchen roll and cloths with a roll of reusable bamboo sheets that are washable and can be used up to eighty-five times. There are twenty sheets on a roll, so each roll avoids at least 1,700 pieces of kitchen roll going in the bin.
- Be frugal – only use one sheet where one sheet will do!

Additional fact

The University of Colorado's Environmental Center reports that: 1 tonne of recycled paper saves 90,849 litres of water; uses 64 per cent less energy and 50 per cent less water to produce; creates 74 per cent less air pollution; saves seventeen trees; and creates five times more jobs than 1 tonne of paper products from virgin wood pulp.

The chemical PFOA which is used among other things to make non-stick pans is to be banned from 2020, with only a few exceptions including safety equipment such as fire extinguishers

SAUCEPAN

Since humans invented the first earthenware cooking vessels, pots and pans have been changing to help us to develop new cooking techniques, from boiling and sautéing to stir frying.

The first metal pots were made of copper and used by the Egyptians, Greeks and Romans to boil water and cook food. Cast iron was introduced as a cheaper option to copper pots in the 15th century when pots were designed with three legs to sit over a fire. The introduction of wood-fired ranges and, later, electric and gas hobs and stoves led to changes in the design of pots and pans as well as the range of materials they could be made from, including stainless steel, ceramic, aluminium, glass and non-stick.

THE IMPACTS

Making a pot from steel, iron, copper or aluminium means that the metal needs to be mined and smelted. Mining has obvious effects on the environment as does smelting, which is highly energy intensive. Aluminium production uses more energy and has a higher carbon footprint than steel or copper production and the lightweight nature of this metal means it may not last as long.

There have also been some health concerns about cooking acidic food in aluminium pots – due to fears that aluminium will leach into the food as it cooks and could contribute to Alzheimer's disease. However, a link has not yet been established.

Stainless steel has a slightly higher carbon footprint than steel as it requires more energy to make it, but stainless-steel pots last a long time and are durable so they can be a good investment. Stainless steel does not conduct heat well, so pots made from it tend to have a copper or aluminium rim in the base.

Copper pots conduct heat really well and are long lasting but, on the downside, tend to be more expensive. Non-metal pots and pans are made from ceramic and glass, the raw materials for which are sand and clay. Ceramic cookware is not commonly recycled and the glass used in cookware is different to that used in bottles and does not tend to be accepted for recycling at bottle banks or in household recycling.

WHAT YOU CAN DO

- Opt for cast-iron or stainless-steel pans when buying new. They have a lower carbon footprint and are durable. Cheaper pans with handles that fall off and non-stick that peels and chips are a poor investment.
- Buy generalist pans that can have many uses instead of having a separate pan for each cooking task. Avoiding plastic handles means a saucepan can go from the hob to the oven and vice versa.
- Put the lid on your saucepan when cooking - roughly 60 per cent more energy is needed to boil a pan of potatoes without a lid, and it takes twice as long.
- Best for the environment to avoid PFTE non-stick pans if you can as this reduces the need for chemicals like PFOA in saucepan manufacturing.
- Look for eco-friendly non-stick pans that use ceramic non-stick coatings.
- Accept hand-me-down cast-iron pots and pans from friends and relatives. Seek them out at car boots sales and in flea markets.
- Pass on old pots and pans as they cannot go in household recycling bins. Metal pans might be accepted by a local scrap metal merchant or can be donated to a charity shop.
- Dispose of old PTFE non-stick pans carefully. Disposal is problematic as they may contain toxins. There are some companies that sandblast pans to remove the PTFE, but this is not a commonly available service.

BAN ON CHEMICAL USED IN NON-STICK PANS

While a non-stick frying pan may produce great pancakes, the substance that gives its non-stick ability has significant impacts on the environment. The non-stick coating was invented in 1938 and when applied to a saucepan prevents food from sticking making it easier to clean.

Polytetrafluoroethylene (PTFE), which is used to make non-stick pans, is made of a mix of perfluoroalkyl substances (PFAS). Certain PFAS are known to accumulate in the environment and in humans and water, and there is evidence that exposure to PFAS can lead to health problems. One of these chemicals, PFOA, is dangerous for the environment and is associated with negative health effects. The United Nations Stockholm Convention on Persistent Organic Pollutants decided to phase out the use of PFOA in May 2019, effective in most countries by 2020.

More than 480 billion plastic drinking bottles were sold worldwide in 2016, up from about 300 billion a decade ago. Sales are expected to continue to grow – to over 570 billion by 2021.

PLASTIC WATER BOTTLE

Plastic water bottles are one of the most omnipresent single-use plastics. From city streets and refugee camps to beachside bars and school canteens, plastic bottles are everywhere. About half of all the plastic bottles used in 2018 in England were water bottles.

While bottled mineral water first came onto the market in Europe in the 1700s, it wasn't until the 1970s that water was packaged in plastic bottles and sold alongside soft drinks by companies such as Coca Cola and Pepsi. The introduction of bottled water into the market in the 1980s was greeted with scepticism – would people pay for bottles of water when you could get water free from any tap? The first part of the marketing plan was merely to have the concept of bottled water accepted – next was to build a market for it.

Convenience is a key driver of sales of bottled water along with the need in many parts of the world to buy clean water where there are untreated or contaminated water supplies. Very worryingly, less than half of the plastic water bottles bought in 2016 were collected for recycling and just 7 per cent of those collected were turned into new bottles.

Plastic water bottles are made from polyethylene terephthalate – or PET for short – a plastic belonging to the polyester family. The main ingredients are crude oil and natural gas, and various additives give the plastic flexibility and clarity. PET is recyclable after washing and shredding into pellets for reuse. Plastic bottles can be recycled into T-shirts, sweaters, fleece jackets (see also pages 108–109), insulation for jackets and sleeping bags and carpets.

A 2017 study by Greenpeace found that six of the largest soft drinks companies, excluding

WHO'S DRINKING BOTTLED WATER?

In the UK, 38.5 million plastic drinks bottles are used every day of the year – only just over half of them make it to recycling, while more than 16 million are put into landfill, burnt or escape into the environment.

In 2015, consumers in China purchased 68.4 billion bottles of water and within a year this had increased by 5.4 billion bottles.

A global look at bottled water consumption

Check out the figures below to discover which populations drink the most bottled water. Expressed as litres per capita, the range in numbers is huge and the variety quite suprising.

Annual consumption of bottled water:
(litres per capita)

Country	Litres per capita
Germany	184
US	171
Indonesia	97
France	77.7
New Zealand	60.9
Ireland	57
UK	44.7
Australia	29.2
Canada	27.6
India	17.2
Sweden	9.3
South Africa	1.4

Coca-Cola, used a combined average of just 6.6 per cent recycled plastic globally. The study also reported that drinks companies appear to have prioritised reducing the weight of plastic bottles (to cut the costs and impact of the production and transport), over their recycling or reuse. But reducing the weight of bottles does not solve the litter and pollution problems.

It is possible to make a bottle from 100 per cent recycled PET, but this is yet to become common practice as the transparency of the bottle is affected. Some plastic from water bottles has been recycled into new bottles – in 2018 and 2019 water bottles made from 100 per cent recycled plastic started appearing on shelves in Europe.

THE IMPACTS

Plastic water bottles are one of the top three pieces of rubbish found in the ocean, after cigarette butts and food wrappers. Over 1.5 million bottles were collected during an international coastal clean-up day led by the Ocean Conservancy in 100 countries in September 2017.

In order to reduce litter and capture the resources in plastic bottles for reuse, deposit schemes are being put in place to create incentives to recycle – these are operational in forty countries and twenty-one US states. A small levy is added to the price of the drink, which is refunded to the customer when they take the bottle back for recycling.

This system is being trialled in some supermarkets in the UK and the US. In Norway, Germany, Lithuania and New South Wales in Australia, reverse vending machines take back the bottles and pay the user in cash.

The production and transport of plastic bottles of water requires the use of a lot of energy. A study published in 2009 found that, even back then when there were fewer bottles made, the long-distance transport of water bottles was comparable to, or even larger than, the energy needed to produce the bottle. The authors estimated that producing bottled water requires 2,000 times the energy of producing tap water and that most of the energy used to produce bottled water has come from fossil fuels and contributed to climate change.

WHAT A WASTE OF WATER!

It takes approximately 3 litres of water to produce 1 litre of bottled water – largely due to the water used to produce the packaging. This translates to more than 100 billion litres of water being consumed every year in the production of bottled water. When compared with current drinking water shortages in the world (for instance, only 24 per cent of sub-Saharan Africa has access to safe drinking water) and the increase in water scarcity associated with climate change, the production of bottled water is far from sustainable.

The World Health Organization's review into the potential risks of plastic in bottled drinking water in 2018 found that more than 90 per cent contained tiny pieces of plastic or microplastics. The study, which analysed 259 bottles of water from nineteen locations in nine countries across eleven different brands, found an average of 325 plastic particles for every litre of water. Only seventeen of the bottles analysed contained no microplastics. A previous study had also found high levels of microplastics in tap water, but the levels of plastic in bottled water were twice that of tap water.

Another report by the WHO published in August 2019 found that there was no evidence so far that microplastics in drinking water are harmful to human health, but warns there isn't enough information to draw a firm conclusion and that more research is needed.

Another health concern linked to plastic bottles is BPA, or bisphenol A. It's a common building block in resins and some types of plastic. It's what's known as an endocrine-disrupting compound, meaning it acts as if a hormone or interferes with normal hormone functions. Research has found that BPA can leach from plastic bottles and food containers into foods and drinks, especially if exposed to heat. The US Centers for Disease Control and Prevention's 2003–2004 National Health and Nutrition Examination Survey found that 93 per cent of people surveyed (from a total of 2,517) had detectable levels of BPA in their urine. This is why more and more bottles now include 'BPA-free' labels.

WHAT YOU CAN DO

- Enjoy tap water. If you live somewhere with clean tap water – drink it. Plus, you'll save money.
- Buy yourself a reusable water bottle and keep it handy so that you have water when you travel and are at school, work or exercising. Stainless-steel water bottles keep your water cool and tasting good. BPA- and other toxin-free reusable plastic water bottles are also an option – look out for brands that will take damaged or old bottles back for recycling.
- Look out for water fountains and hydration refill stations in workplaces, towns, airports and train stations so that you can refill your water bottle when you are away from home.
- Invest in a soda stream and making your own fizzy water rather than having to buy bottled versions in plastic.
- Glass bottles are a better choice than plastic if you decide to buy bottled water as glass is easier to recycle.
- Reuse the plastic water bottles you already have – they are really durable so if you have one reuse it before you recycle it.
- Recycle, always. If you're out and about, look for a recycling bin or if none is obvious, bring the bottle home and pop in your home recycling bin. Plastic bottle lids are recyclable, too, so be sure to put the lid back on empty bottles.
- Repurpose empty bottles – use them as plant pots, bird feeders and paint brush holders.
- Go large. If you visiting a country where the water isn't safe to drink from the tap, buy a big (8-litre) bottle of water and refill a smaller bottle from that to reduce your plastic waste.
- Take water purifying tablets with you if you're holidaying but the local water isn't safe to drink.

Additional facts

The average weight of a 480g PET plastic bottle was 48 per cent lighter in 2014 than a similar bottle in 2000, saving 2.8 billion kg of PET resin since 2000.

In Norway, where there is a plastic bottle deposit scheme, 95 per cent of all plastic bottles are recycled, compared with only 57 per cent in England.

Globally we use an estimated 5 trillion plastic bags a year. That equates to 700 bags per person per year and a large proportion of these are bin bags.

BIN BAG

Bin bags are designed to line rubbish bins so that the containers themselves don't get dirty and smell, and to make emptying them easier and less messy. Designed to be disposable, they are used for an hour, a day or a week to collect and transport rubbish before they, too, become waste along with their contents.

We have only been using plastic bin bags since the 1950s, yet they are now widely used in bins in offices, homes and restaurants as well as at events and on the street. In recent decades, a wide variety of these bags has been produced ranging from small, transparent bags designed to line office waste-paper baskets, to thick, black and green plastic bags for gardening waste.

It is estimated that the global market in bin bags was US$ 1.4 million in 2017, but this is projected to rise to US$2.3 million by 2024, due largely to growth in the emerging economies of India and China.

In many parts of the world it is commonplace to line every bin in the home, office or hotel with a plastic bag. While this arguably makes it more convenient for the person emptying the bins, it means multiple bags are being used and the impact on the environment is huge, especially if these smaller bags are then put into a larger bin bag when collected up to be thrown out.

The paradox of the bin bag is that it should be strong enough to hold waste without leaking or splitting, which makes plastic an excellent material to produce it from; yet the durability and longevity of plastic is a mismatch with a bin bag's short-term use. Bin bags are not designed to be recycled; by their very nature, they are designed to be thrown away, and at present there are few facilities that can recycle these soft plastics.

THE IMPACTS

Plastic bin bags have several significant environmental impacts. Most of them are produced from virgin plastic made from crude oil, which is a finite natural resource. They also last a long time – potentially for up to 1,000 years. Most landfill sites are anaerobic (without air); when deprived of oxygen and micro-organisms, organic waste – including the contents of bin bags – decomposes very slowly, releasing methane gas and dangerous leachates (toxic liquids) in the process (for more on what happens in landfill, see pages 46–47).

Out in the environment, plastic bin bags pose a threat to wildlife. In the ocean, plastic bags look like jellyfish and are mistakenly eaten by turtles, whales and sea birds. In 2018 an autopsy on a sperm whale in Indonesia found that it had ingested 115 drinking cups, twenty-five plastic bags, plastic bottles, two flip-flops and a bag of more than 1,000 pieces of string.

When marine animals swallow plastic they feel full and stop eating, and in seabirds a stomach filled with waste can weigh them down and impair their ability to fly. As plastic bags break down in the environment, they tear into smaller and smaller pieces to become microplastics, which are mistakenly eaten by fish and other marine life, ultimately making their way back to us in our food.

WHAT YOU CAN DO

- Count how many bin bags you currently use per week. Multiply that number by fifty-two weeks to get a sense of the annual cost saving and positive impact on the environment you'll make through reducing the number of bags you're using.
- Wash the bin after emptying instead of lining bins with plastic bags. Not putting a bin liner in a kitchen bin may sound like a recipe for a smelly bin, but if all your food waste is going in your brown bin or compost and you are recycling glass, plastic and cans, all that is really left in your kitchen bin is soft plastic. So, all kinds of plastic bags and food wrappings as well as plastic films. Give it a try for a week and see how you go.
- Assess which bins in the home (and office) collect dry, clean waste such as paper and wrapping and so don't need to be lined with a plastic bag. Liners may be unavoidable for bathroom bins, for nappies or sanitary waste. But some bathroom waste is recyclable – shampoo bottles or toilet rolls. Separate these items out for recycling – you will reduce waste and need fewer bags as they won't fill up so quickly.
- Never bag up recycling. Most local councils provide a separate curbside bin for recycling which you can put your recyclables into directly. Plastic bags get tangled in the conveyor belts at recycling centres and so recycling sites prefer not to have them. If your council provides a bag to put your recycling in, it is because they have a process for de-bagging it at the recycling centre.
- Your brown or organic waste bin doesn't need a liner – just tip food and garden waste straight in or drop in the bags you use to line your kitchen caddy. If it gets a little smelly hose it out and add some bicarbonate of soda (see also page 69). It's best to only use the compostable bags your council provides or recommends.
- Reuse plastic or paper shopping bags as bin liners – much better than buying new bags.
- At times when you absolutely need a bin bag, look for bags made from recycled plastic – a better choice than virgin plastic.

A LOOK AT WASTE

Every day we produce waste, yet we pay little attention to it unless there is nowhere to dispose of it or our regular waste collection services are disrupted. One or two weeks without a bin collection and most of us start to notice just how much waste we accumulate. In 2016/17 the average person in the UK produced 412kg of waste (that's about five to six times the weight of an average adult), up from 406kg per person in 2015. Even though we are recycling more, every year we produce ever more waste because we are buying and consuming more stuff, which manufacturers and retailers continue to overpackage.

In 2016/17, 45 per cent of household waste in the UK was recycled, well on the way to reaching the 50 per cent by 2020 target set by the European Union. This is a positive step as it keeps glass, plastic, paper and aluminium out of landfill and incineration and means materials are reused, which saves energy and natural resources. The rubbish left over, though, is incinerated or goes to landfill.

In the UK in 2016/17, 10.2 million metric tonnes of waste were incinerated and 4.1 million metric tonnes went to landfill. More UK waste goes for recycling nowadays, but there is a lot of room for improvement. There were 604 landfills in the UK in 2016 (both in use and closed but still managed) and seventy-eight incinerators.

WHAT IS A LANDFILL?

A landfill site is literally a hole in the ground where rubbish is dumped. Modern landfills are carefully located in terms of their geology, groundwater, accessibility and community approval. The hole is lined with clay and an impermeable membrane so that no liquid, called leachate, can escape. Drains built into the landfill capture the leachate and pump it to wastewater pools located beside the landfill, before it is treated as wastewater and released. A groundwater monitoring system detects any leaks and allows steps to be taken to protect the environment from potential toxins.

At a landfill site, every day the waste is levelled out and covered with clay to keep down the smell and dust. The methane gas generated by the waste is captured and burned or can be diverted and used as biogas to create heat, fuel or electricity (see also page 174).

When the hole is deemed full, it is covered in the same impermeable membrane and then soil. It's possible to landscape such sites to create a new amenity such as a park. The groundwater will continue to be monitored and the methane gas captured until biodegradation and the production of the gas ceases.

Old landfills built before modern regulations don't have a liner, so the leachate can contaminate groundwater and the methane gas that has been escaping from them for years will continue to do so.

THE RECYCLING SITUATION

In the past, materials such as plastic collected for recycling were sent to China for processing and often ended up being burned or dumped. However, in 2018 China announced it was no longer accepting waste from other countries, a statement that's shaken up recycling systems around the world.

Plastic from the UK is now being recycled in Malaysia, Turkey, Poland, Indonesia and the Netherlands. Malaysia has seen the biggest increase in UK plastic imports and there are concerns about its capacity to recycle all the plastic it receives as well as fears about illegal dumping.

Glass, paper and aluminium and other metals are recycled in the UK, and where volumes are large may also be exported to the EU for recycling. Ultimately the aim must be for every country to be responsible for its own waste and to have systems locally to reuse, repurpose and recycle.

WHAT HAPPENS IN A LANDFILL?

Landfill sites are what's known as anaerobic or 'without oxygen'. Because such piles of rubbish are pressed down on one another, there is effectively no air. No oxygen means no micro-organisms to help the decomposition of the waste, so that happens but only very slowly. If materials such as metal and glass make it into landfill, they don't decompose, they just sit there; the same goes for plastic (see also page 134). Even food, paper and garden waste, which will biodegrade if placed in compost bins or left in the open, break down extremely slowly in landfill and they release methane gas in the process.

Methane is a powerful greenhouse gas – it's twenty-eight times more potent than carbon dioxide, even though it stays in the atmosphere for a comparatively shorter amount of time (twelve years versus 100–300 years for CO_2). Although modern landfills capture methane gas and burn it, only a small proportion of sites use it to generate electricity and heat.

FOOD WASTE AND CLIMATE CHANGE

Food waste alone is responsible for between 8 and 11 per cent of global warming – in fact, if food waste was a country it would be the third-largest emitter globally, coming just after China and the US (see also page 22).

Sending food to landfill is the worst way of disposing of it and avoiding food waste ranks third out of a list of 100 climate solutions collated by Project Drawdown, a research project to identify the top 100 solutions to climate change. More and more councils and municipalities are now providing kerbside food waste collections, which divert organic waste from landfill where it produces methane and instead converts the waste into valuable compost. In fact, harmless black soldier flies are being used to decompose food waste more quickly and to produce a fertiliser and fish food as a by-product.

What's more, we are also running out of space for landfill. In some places there just isn't physical space for these sites, but it is also because nobody wants to live near a landfill. Incinerating waste rather than dumping it is becoming more popular as the energy generated can be used to create local heat and power, so that the waste essentially becomes a fuel.

REDUCE, REUSE, RECYCLE

What's the answer? Choosing between landfill and incineration and trying to decide which is the least bad option is no solution. The answer to our waste problem is to produce less waste in the first place (reduce), to reuse and repair everything we can (reuse), to recycle what can't be reused or repurposed – everything from plastic to glass, paper and metal (recycle) – and to compost all food waste. Then and only then, should we resort to incineration or landfill. Neither option is ideal, both have environmental impacts and as an individual you don't get to choose where your waste ends up – it is decided by your local authority.

Ideally we should get to a place where nothing is created that can't be repurposed, recycled or composted – in this way we can achieve zero waste.

More and more zero-waste shops are opening that allow you to buy food, cleaning products and cosmetics without packaging, greatly reducing household waste. Using your own containers for takeaway lunches can also help to reduce the estimated 11 billion items of packaging waste generated each year from takeaway lunches in the UK.

SHOULD WE BE BURNING RUBBISH?

Incineration burns waste at high temperatures in huge furnaces to reduce the solids burned by up to 95 per cent and producing ash in the process, which is collected. The gases produced are passed through filters and cleaners to extract any toxins before being released into the air. The toxins created by burning plastics, in particular, are hazardous and are concentrated in the ash and in the filters, which require specialist disposal. Despite efforts to reduce pollution from incineration, concerns remain about the release of toxins and pollutants such as dioxins. Dioxins are bad for the environment and dangerous for human health.

The most efficient incinerators convert the heat they produce to provide district heating or to generate electricity using steam to drive turbines and are known as waste-to-energy facilities. A new waste-to-energy plant in Addis Ababa in Ethiopia is giving employment to former waste pickers, making bricks from the detoxified ash by-product. Such plants can provide constant and predictable electricity to complement more intermittent renewable sources.

LABEL CONFUSION

The different labels and symbols that are used to tell us what we can and can't do with regard to packaging and waste can be extremely confusing. For example, the green dot (a circle of two intertwined arrows) that tells us that the manufacturer of the product has contributed to the cost of packaging recycling is frequently misunderstood; in fact, half the people surveyed by *Which!* in the UK in 2018 thought the arrow meant that the package was recyclable. Things are changing all the time as technology and recycling services develop – so keeping up to date can be hard. As a rule of thumb:

- food waste should be composted
- clean and dry paper, cardboard, tin cans, drinks cans and glass should be recycled
- hard plastics, such as the punnets fruit, veg and meat come in should be recycled (except for black trays that can't be distinguished from conveyor belts in recycling centres)
- soft plastics, such as the bags bread, pasta and rice come in, anything dirty (nappies, oily pizza boxes) and anything you are not sure about should go in the rubbish bin.

IS COMPOSTING THE FUTURE?

Given the fact that so many things are not recyclable, such as dirty paper plates and disposable coffee cups, ever more single-use items are now designed to be compostable instead. Compostable means that the material will biodegrade in an industrial composter within nine to twelve weeks. It doesn't mean it is suitable for home composting, which works at much lower temperatures.

- Industrial composting – Temperatures in an industrial composter reach up to 50–60°C; a home compost bin typically reaches 20–40°C. Industrial composters are big warehouses where the waste is made into a giant compost heap that is turned regularly to ensure aeration; aeration can also be achieved by feeding air into the compost from below. The temperature, oxygen levels and humidity are engineered to be perfect for microbes to break down the waste into a nutrient-rich compost that can be sold for use in gardens and for landscaping. When you put your food waste or organic bin out for collection it ends up at an industrial composter.
- Home composting – A compost bin in the garden works well for vegetable and fruit peelings but not for compostable cups, packaging and plates. So, keep compostable packaging and cups out of your home composter. As waste decomposes, carbon dioxide and water are released and a biomass, or compost, remains that can be used in the garden.

Although many things are referred to as 'biodegradable', meaning it will break down in time, that could mean anything from a few months to 1,000 years. So, something labelled as biodegradable is not compostable – compostable items have to be labelled clearly as such. Do make sure anything compostable gets composted – or the extra cost and effort that went into providing a compostable container is literally wasted.

THE UTILITY ROOM

Over 840 million washing machines are in use worldwide. Each machine produces, on average, 0.7kg of CO2e, the equivalent of driving 1.7 miles (2.7km), per load of laundry washed at 40°C. If you do four loads a week that is equivalent to a 7-mile (11-km) journey in a standard petrol car.

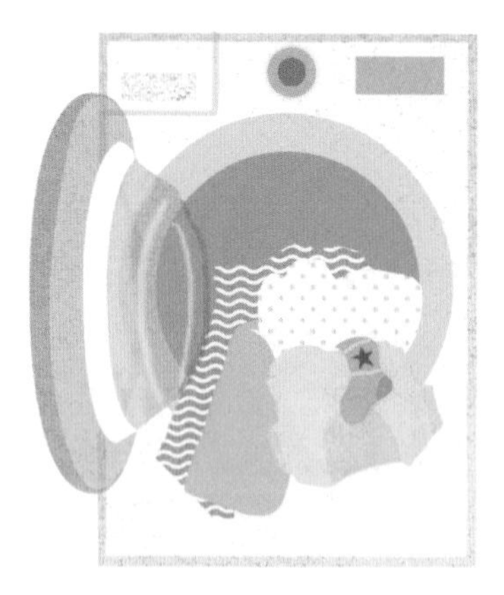

WASHING MACHINE

As recently as a couple of generations ago, all clothes washing was done by hand. You may remember seeing a wooden washboard in your grandparent's home or the arrival of the first washing machine (which sometimes flooded the house if left unwatched). The washing machine is a relatively new invention and it is one that has revolutionised the lives of women in particular.

Hans Rosling, a Swedish statistician, calls the washing machine 'the greatest invention of the Industrial Revolution'; and Ha-Joon Chang, an economist at Cambridge University, claims that 'the washing machine changed the world more than the internet'.

But first let's look at the history of this revolutionary machine. In the past, washing clothes happened outside the home. The fullones were the launderers of ancient Rome, washing white togas made of wool and linen. Detergent and soap hadn't been invented so clothes were washed in water with plenty of animal or human urine, which acted as an alkali, helping the dirt separate from the clothes. Men and boys stamped up and down with their feet in vats to clean the clothes before they were dried, brushed (with a hedgehog pelt) and sometimes smoked with sulphur to whiten the cloth.

Soap came along in medieval times but only for the 'well to do'. By the 18th century soap was common but still used sparingly, so was kept for stain removal and for 'good' clothes rather than for cleaning the whole wash.

The first washing machine was a type of cage with wooden rods and a handle for turning it, credited to Henry Sidgier of Great Britain in 1782. This basic design eventually led to the invention of the revolving drum in 1851. The electric washing machine was first sold in the US in the early 1900s. By the 1970s, 65 per cent of UK households had a washing machine, rising to 90 per cent in the 1990s and 97 per cent by 2016.

Yet, the washing machine is still a luxury item. Only 2 billion of the total world population of 7.7 billion can afford a washing machine. That leaves the rest of the world, over 5 billion people, washing their clothes by hand, in a stream or with water pulled from a well and carried home. This is all unpaid work, largely carried out by women, taking up time that could be spent with their children, in paid employment or developing new skills. In Peru, a study conducted in a slum with 30,000 residents found that doing laundry took up six hours a day three or more times a week. Compare this to the UK where an average person spends 102 minutes per week washing clothes and hanging them out to dry (2017).

THE IMPACTS

Washing clothes consumes energy and it is the running costs of a washing machine that contribute most to the carbon footprint of clothes washing; plus there's the impact of the detergents used (see also page 56). Studies into the total cost of owning a washing machine show that over 60 per cent of the costs are associated with the energy and products consumed in using the machine (such as detergents), while 23 per cent of the cost is associated with the purchase of the machine. Research by *Which!* shows that the energy costs of running a washing machine vary between £140 and £250 a year based on average washing cycles. *Which!* estimates that you could pay £80 more a year in energy costs if you buy one of the least-efficient models. So, buying a more energy-efficient model, even if it is slightly more expensive, can save you money in the long-run, especially if you consider most of us will use the same washing machine for six to eight years.

People in the US do more loads of laundry than anywhere else – 289 washes per household per year. However, energy consumption per wash cycle is higher in Europe than in the US and Australia, most likely because the machines are washing at a higher temperature. However, the water consumption associated with washing clothes is highest in Japan and the US, with European washing machines using least water. This can be linked to the design of the machine: front-loading machines in the European Union consume much less water per wash cycle than top-loading machines (favoured in the US), which often run exclusively on cold water, and hence have lower electricity costs.

The impact of microplastics on the world's oceans has deservedly received growing public attention in the last few years. A study at the University of Plymouth in 2016 looked at the release of synthetic, microplastic fibres from domestic washing machines and found that a 6kg wash of clothes can result in anything between 137,951 fibres (for polyester cotton clothes) to 728,789 fibres (for acrylic clothes) being released into waterways. These fibres are so small they pass right through sewage treatment screens and are released into aquatic environments where they accumulate in the

HOW WILL WE WASH CLOTHES IN THE FUTURE?

The washing machine of the future is already here. It's currently being trialled in commercial settings, such as hotels, with great success.

The Xeros washing machine uses beads called XOrbs to reduce the water they use per wash by 80 per cent. Instead of using water to mechanically remove dirt and stains, beads held within the machines clean the clothes using a lot less water and with a lot less impact on your washing – so, clothes keep their colour and texture longer.

A hotel and spa in the UK that has replaced two normal washing machines with these new Xeros machines is saving 2.5 million litres of water each year – more than an Olympic-sized swimming pool!

The Xeros washing machines use only half the electricity of regular washing machines, so running costs and greenhouse gas emissions are lower. A hotel in the US took less than one year to recoup the cost of investing in the Xeros machine from the cost savings due to lower water and energy bills.

And, even better, the latest prototype for home use has a filter to catch the microfibres (tiny plastic fibres) that are released by synthetic clothes when they are washed. The filter is cleaned out every few months and prevents microfibres from entering waterways and the ocean.

Look out for these washing machines, and others like them, when they hit the shelves in the coming months and years.

sediment or are ingested by wildlife. This means that a city the size of Berlin may be responsible for releasing the equivalent of 540,000 plastic shopping bags' worth of microplastic fibres into the ocean every day.

Washing machines are currently designed to last for ten years and to do 250 cycles per year. Making washing machines last longer reduces their carbon footprint. For example, replacing five 2,000 cycle machines with one 10,000 cycle machine yields almost 180kg of steel savings and more than 2.5 tonnes of CO_2e savings.

In the future we will lease more durable washing machines, with a longer life expectancy, making it worthwhile to repair and refurbish the machines and reducing the overall carbon footprint and the cost to the consumer (26–38 per cent cheaper to lease than own) of washing clothes.

WHAT YOU CAN DO

- Wash your clothes less often. Check they are actually dirty before putting them in the wash basket. Spot-clean stains to avoid washing the whole garment. Try airing your clothes to get rid of odours, especially environmental odours such as smoke or cooking smells that will dissipate easily during a spell in the outdoors. Washing your clothes less also means that they will last longer and the carbon footprint of the item of clothing will be reduced (see also page 112–113).
- Do full loads of washing. Get value for money and make sure the load is full but not over full; stuffing more items into the machine will simply mean that you will end up having to wash some items again.
- Wash clothes on cold or at 30°C. Most liquid washing detergents work well at low temperatures these days and turning down the temperature is an extremely effective way to reduce your energy costs. Switching from the 60°C to the 40°C setting will cut your washing costs by about a third and most clothes can be washed at 30°C or even on cold.
- Learn how best to use your machine; it might sound obvious but lots of people never read the manual or discover which modes are the most efficient. The latest washing machines can even calculate how much detergent and water to use based on the load to further increase efficiency.
- Buy the most energy- and water-efficient washing machine you can afford. You will save money in the long run from a more expensive machine that costs less to run. A high-end efficient machine costs roughly 10 pence per washing cycle, while low-end machines cost 21 pence per cycle. Look out for A+++ rated machines.
- Be sure to recycle your old washing machine; it contains 30–40kg of valuable steel, plus electronics and plastics that should not end up in a landfill. There are specific areas for electric waste at recycling centres and many retailers offer a pick-up service for old appliances when you buy a new one.
- Switch to a 100 per cent renewable electricity provider to reduce the carbon footprint of the energy used to power your machine.
- Run your machine at night if you have a reduced night-time energy rate; get in touch with your electricity supplier to check rates.

A detergent concentration of only 2 parts per million (ppm) can interfere with how fish absorb chemicals, increasingly their vulnerability to pesticides, for example. Most fish will die when detergent concentrations approach 15ppm, even concentrations as low as 5ppm will kill fish eggs.

LAUNDRY DETERGENT

Before the advent of laundry detergent, clothes were washed with water and urine. Early soaps were made from animal fat boiled up with the ashes from a fire (see also pages 52 and 124).

Detergents are synthetic, water-soluble cleansing agents and, unlike soap, don't form a scum with the salts contained in hard water. They have many ingredients: surfactants, which lower the surface tension of water to enhance the wetting, foaming, dispersing and emulsifying; phosphates, which soften hard water and help suspend any dirt in the water; optical brighteners, which make the items reflect light; enzymes in biological detergents, which help to break up and remove food and other stains; and perfumes, which are added to make it smell nice.

THE IMPACTS

All detergents have environmental impacts, even biodegradable ones if they reach sufficient concentrations. Detergents don't discriminate from water in your washing machine or out in the natural world – they act in the same way. Their many ingredients act to attack the natural oils in the mucus membranes of fish, stopping their gills from working properly and increasing their risk of exposure to bacteria, parasites and pesticides. The surfactant ingredients in detergents produce what are called 'endocrine disruptors', which can affect an animal's hormonal balance and reduce fertility; yet other ingredients can be toxic and persist in the environment.

In the 1960s and 1970s scientists started to understand just how phosphates affected water quality and contributed to the phenomenon of eutrophication (when chemical-rich water ends up in environmental water). Excessive levels of phosphates in water lead to algal blooms that release toxins and deplete oxygen in waterways, killing aquatic life. To address this issue, Ecover started producing phosphate-free laundry detergent in the 1980s, replacing the phosphate with a microporous mineral called zeolite. Established commercial manufacturers were slower to follow, phasing out phosphates in the US in 1990s and then in Europe in response to regulations agreed in 2013. Targets to end the use of phosphates in other markets around the world are more recent.

WHAT YOU CAN DO

- Use a phosphate-free detergent with plant-based surfactants that have minimal aquatic toxicity and also biodegrade quickly (see also page 37 for easy to see descriptions on labels). If in doubt about brands, choose a well-established eco-brand or do some research online.

PLANETARY BOUNDARIES

The Stockholm Resilience Centre looks at the sustainability of what's known as the biosphere – the worldwide sum of all ecosystems, sometimes called the zone of life. Phosphorus and nitrogen together are just one of nine crucial cycles that make the planet work. Human activities have skewed the levels of these chemicals, which go on to have wide-ranging effects on nature. Reducing the use of phosphates in detergents can lessen the risk the planet is currently exposed to – a small contribution to guarantee future life on Earth for all.

- Measure your detergent carefully and always use the correct amount.
- Choose laundry liquid as it gives better results than powder for cool washes; that said, it is heavier to transport as it contains more water. So, the more concentrated versions are best. Liquids tend to be packaged in plastic containers (whereas powders come in cardboard boxes) – so make sure you refill yours and ultimately recycle it. Look out for laundry liquids in recycled plastic containers.
- Recycle the cardboard packaging of powdered detergent; be sure to remove any plastic handles and make sure the box is completely empty. The ingredients in powder are more stable, giving it a longer shelf life and reducing the need for preservatives.
- Avoid laundry detergent tabs and capsules completely as they contain unnecessary colouring agents and other ingredients in their soluble packages.
- Refill. Ever more shops and supermarkets are offering refill stations for laundry detergent. That way you can use the same container for years.
- Discover eco eggs – these laundry eggs remove the need for detergent completely. Two types of natural mineral pellets inside the reusable egg work together to clean clothes for up to 720 washes.
- Avoid using a fabric softener; it introduces unnecessary chemicals into your washing, to your skin and into the aquatic environment.
- Keep your washing machine clean to allow detergent to work more effectively. Research online to find a homemade solution for this based on bicarbonate of soda and vinegar (no need for a specialist cleaning product); see also page 69 and 247).

In 2018, 58 per cent of UK households owned a tumble dryer – that's a massive 15.7 million machines.

TUMBLE DRYER

If you live in a wet or damp climate or without access to an outdoor space to line-dry clothes, a tumble dryer can be a life saver. But these handy machines guzzle energy so it's time to revisit how we use them.

Tumble dryers are a relatively modern invention but their origins go back some 200 years. The first version was a metal drum with holes that was rotated over an open fire. Invented in 1799 by Monsieur M. Pochons (a Frenchman), it dried the clothes but left them smelling of smoke. Despite its shortcomings, a ventilator system comprising of a metal drum with holes is the basis of the tumble dryer to this day.

The first tumble dryers for home use were available in the early 1900s, but came with a hefty price tag. It wasn't until the late 1930s that gas and electric tumble dryers became more affordable and competition between manufacturers started to drive down the price.

Domestic tumble dryers are responsible for up to 10 per cent of total residential energy use in developed countries. So, there is a growing trend in Europe towards more efficient designs. Heat-pump tumble dryers, for instance, work by reheating the air that circulates in a tumble, pumping heat back through the drum and reducing energy use by up to 60 per cent compared with a conventional tumble dryer.

THE IMPACTS

Tumbler dryers use a lot of energy and if that energy is generated from fossil fuels its production releases greenhouse gases that cause climate change. As a result an A-rated (energy-efficient) tumble dryer used three

times a week generates over 160kg CO_2 per year – that is equivalent to leaving a large-screen LCD TV on continuously for forty days. In addition, tumble dryers are inefficient. Up to 60 per cent of the heat they produce is wasted. So, in fact, tumble dryers are a very expensive way to dry your clothes.

Tumble dryers also accelerate the shedding of microfibres from synthetic clothes. The filters in dryers that collect lint also collect microfibres, and if released into the environment these small plastic fibres contribute to pollution and can enter the food chain.

WHAT YOU CAN DO

- Dry your clothes outdoors on a washing line, ideally. If every household in the UK that owns a tumble dryer dried just one load of washing outside each week (and didn't use their tumble dryers), they would save over 1 million tonnes of CO_2 in a year. Imagine switching to half of all washes or 100 per cent of washes, the savings could be huge.
- Use your tumble dryer as little as possible and consider using your airing cupboard to ensure air-dried clothes are fully dry. For homes without airing cupboards, assign some space to air dry your laundry inside on a clothes airer.
- Use a clothes airer or drying rack. If you don't have a garden with a washing line or are limited to a small balcony, dry your clothes outside on an airer and just lift it indoors at night or if it rains.
- Wash your clothes less! The convenience of washing machines and tumble dryers means that we are doing more and more laundry and washing our clothes too much. Instead, air clothes to get rid of odours or spot-clean stains and spills to reduce the amount of household laundry.

- Spin clothes on a high speed in your washing machine if you have to use a dryer to get rid of as much water as possible before they go in the dryer.
- Recycle your old dryer by taking it to a WEEE collection point (it contains valuable metals that can be used again) and either don't replace it or replace it with an efficient heat-pump model. Heat-pump dryers cost a bit more to buy but the running costs are lower, returning your investment over time.
- Dispose of the lint from a dryer in the rubbish bin, and never wash it down the drain. This means that microfibres don't get released into waterways.
- Don't use dryer sheets. They are single use, contain plastic and produce waste. If you have to use a dryer, try a felt wool dryer ball. They reduce static and the lanolin in the wool makes the clothes softer.

A NOTE ON PEGS

When drying clothes outdoors on a line you need pegs. Plastic pegs weather quickly if left outdoors and become brittle and break into small pieces that can get lost in the grass and work their way into the soil, and possibly into the food chain. So, bring your pegs indoors after use to make them last longer. Pegs cannot be recycled unless you separate the metal from the plastic. It is better to use wooden pegs, ideally made from sustainably sourced wood, or stainless-steel pegs – they're claimed to last a lifetime.

FOOTPRINTS AND LABELS EXPLAINED

Saving the planet really means saving ourselves. Planet Earth can exist quite happily without *Homo sapiens* – but if we make conditions here hostile for our own species we will become extinct. I say this only to make the point that we have a vested interested in making sure the planet is a safe place for our future.

In order for human beings to flourish as a species, we have certain requirements: a stable climate system with temperatures in a relatively narrow range that is neither too hot nor too cold, oxygen to breathe, nutrients from food and fresh water to drink. Earth is sometimes called the Goldilocks planet – because it is 'just right' for *Homo sapiens* and a plethora of other species to thrive.

Now that we realise we're doing harm, we have started to try to account for the impact for the things we do and the goods and services we buy in terms of their carbon footprint or environmental footprint. I'll explain below what these terms mean and why they're used.

WHAT IS A CARBON FOOTPRINT?

Carbon footprint is shorthand for a calculation of the climate change impacts of an object or action, accounting for all the different greenhouse gas emissions and not just carbon dioxide and as a best guess based on available information. To account for the fact that the carbon footprint relates to a number of different gases, not just carbon dioxide, it is expressed as if all those gases were converted to carbon dioxide and so it is called a CO_2 equivalent or CO_2e for short.

So, when we talk about the carbon footprint of a short-haul flight, it is a shorthand for all the emissions generated by the plane on that journey. How the calculations are made varies from one carbon calculator to another – hence the numbers are best used as a guide and relative measure.

HOW CARBON FOOTPRINTS COMPARE

The average carbon footprint of a person in the UK is **5.56 metric tonnes** of CO_2e, while for a person in the US it is **14.95 metric tonnes** as they consume more, live in bigger houses and drive bigger cars. In developing countries emissions per capita are much smaller: in India it is a mere **1.57 metric tonnes** CO_2e per person and in Malawi just **0.1 metric tonnes** CO_2e per person. The countries with the highest per capita emissions are Saudi Arabia, Australia and the United States.

You can calculate your carbon footprint online (see page 251).

WHAT IS AN ENVIRONMENTAL FOOTPRINT?

You may well have read about an environmental – or ecological – footprint. This is a way of calculating all of the environmental impacts associated with a lifestyle, including carbon emissions, water consumption, and the associated land, resources and waste. The result is given in terms such as how many planet Earths are needed to sustain your lifestyle. So, for instance, if your score is '2 planet Earths' you need to reduce your carbon footprint by half – by making change to where you live, the transport you use and the things you consume. There are different methodologies and ways of doing this, however, and no universally accepted formula. Some environmental footprints look at the impact of whole countries. The ecological footprint of the UK, for example, is 4.4 global hectares per person – accounting for the resources, energy and pollution entailed in how the population of the UK lives. Given that globally the Earth can provide only 1.63 hectares per person, it would take 2.7 planet Earths to sustain the world's population if everybody lived and worked as in the UK. By comparison the ecological footprint of India is 1.2 global hectares per person, well within the Earth's capacity.

FROM THE CRADLE TO THE GRAVE

Another assessment tool often used to assess environmental impact is life cycle analysis (or LCA). It tries to capture all of the environmental impacts of an object from cradle to grave and is used by companies to assess the impact of their products. In other words, from the extraction of the raw materials through the manufacturing, use and recycling to disposal. As with a carbon footprint, LCA is a relative measure that allows comparisons and to see where in the life of a product the greatest impacts lie – and therefore where the greatest reductions can be made. There is no exact science to these calculations, but the methodologies are improving. I have referred to the LCA of the objects in this book where that information is available and have used a carbon footprint over an ecological footprint because this information is most readily available for the objects featured.

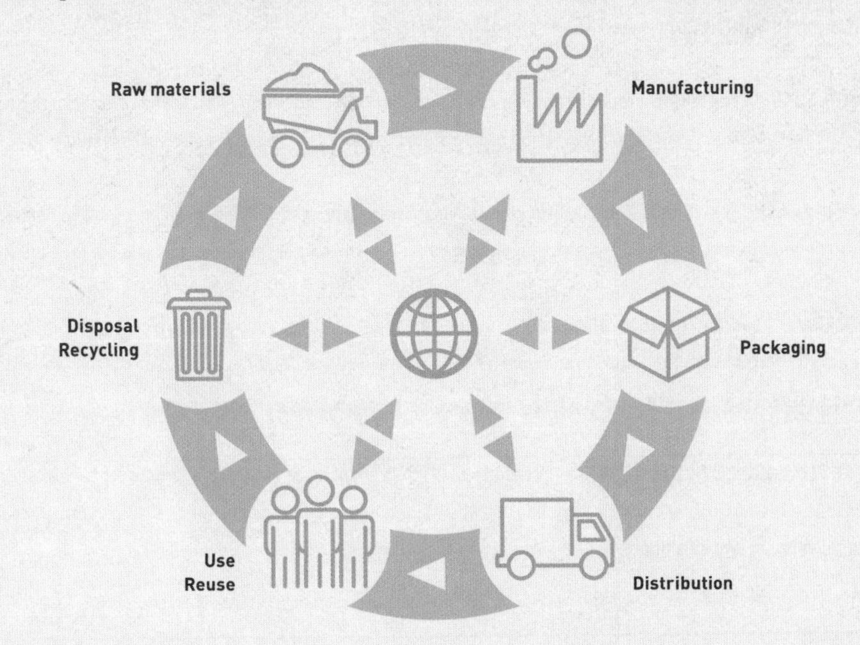

DECIPHERING LABELS

Eco labels are used by manufacturers and other organisations to provide you with assurance that the product you are buying has met certain criteria, which make it more environmentally friendly than conventional products. General statements found on packaging such as 'organic', 'eco-friendly', 'natural', 'plant-based' and 'biodegradable' do not guarantee that the products meet certain criteria and such claims are very often not backed up or verified. So, look for well-established and independent standards and marks such as those in the list below.

B CORPS – Certified B Corporations achieve a minimum verified score on an assessment of the company's impact on its workers, customers, community and environment and submit their report online in the interests of transparency. Globally, over 2,900 companies were B Corps members in 2019.

BETTER COTTON INITIATIVE – products with the BCI label support cotton farmers to use water efficiently, use less chemicals, take care of their soil and biodiversity and receive a decent wage. Brands that carry the mark have to commit to sourcing at least 5 per cent of their cotton from these farmers at the start, rising to 50 per cent after five years. Find this on sheets, towels, T-shirts and jeans.

CERTIFIED CARBON NEUTRAL – certifies companies that calculate their carbon footprint and reduce it to zero by combining efficiency measures in their operations with carbon offsets gained by supporting emissions-reduction projects.

ETHICAL TEA PARTNERSHIP – works to improve the sustainability of tea production, the lives of tea workers and the environment in which tea is produced (see also pages 16–18).

EU ECO-LABEL – awarded to products and services that meet high environmental standards throughout their lifecycle, that reduce emissions and waste and are designed to be reusable, repairable and recyclable.

EU ENERGY LABEL – certifies the energy efficiency of an electrical appliance from A, most energy efficient, to G the least efficient. Look for the label on all electronics sold in the EU. The most efficient appliances are labelled A+++.

FAIRTRADE – used by the food and fashion industries, this ensures decent pay, working conditions and environmental protection. Find this label on coffee, chocolate, bananas and tea as well as cotton clothing (see also page 19).

FOREST STEWARDSHIP COUNCIL (FSC) – products with FSC certification come from forests that are sustainably managed and benefit locally communities. Look for the FSC marks on computer paper, greeting cards, packaging, books, toilet paper and cardboard.

GLOBAL ORGANIC TEXTILE – GOT labels certify clothes as organic that are made from at least 70 per cent organic fibres. Look for this label on reusable nappies, bedclothes, napkins, tablecloths and clothes.

LEAPING BUNNY – a certification programme for cosmetics that are free from animal cruelty. Look for the bunny symbol to know that cosmetics are animal friendly.

PEOPLE FOR THE ETHICAL TREATMENT OF ANIMALS – this animal rights organisation hosts a database of companies that are animal cruelty free and vegan. The companies on the list have PETA's statement of assurance or provide a statement verifying that they do not test on animals.

RAINFOREST ALLIANCE – the seal is awarded to farms, forests and businesses that meet rigorous environmental and social standards. Look for the tree frog symbol on coffee, tea and bananas (see also page 19).

SOIL ASSOCIATION ORGANIC – the world's first organic standard that certifies food as organic as well as meeting criteria relating to animal welfare, human health and environmental protection. Beyond food this label is also awarded to beauty products, cotton for clothing and forests and wood products.

THE CARBON TRUST STANDARD – awarded to organisations that take a best-practice approach to measuring and managing their environmental impacts, commit to reducing these impacts year-on-year and to improving their efficiency over time. Companies with this standard range from banks and supermarkets to the government of Scotland and food companies.

THE GOOD SHOPPING GUIDE'S ETHICAL COMPANY AWARD – the Good Shopping Guide rates UK companies in terms of their commitments to animal welfare, human rights and the environment and gives the Ethical Award to companies that meet their standards.

THE RESPONSIBLE WOOL STANDARD – a voluntary global standard that addresses the welfare of sheep and of the land they graze on. The RWS certification ensures that wool from certified farms is tracked. Look for the RWS mark on wool clothing and blankets.

THE SUSTAINABLE FIBRE ALLIANCE – a global sustainability standard for cashmere production to preserve and restore grasslands, ensure animal welfare and secure livelihoods. Much of the world's sustainable cashmere is from Mongolia and the Alliance supports herders there to graze their cashmere goats and manage their resources sustainably.

About 45 million vacuum cleaners are sold each year in Europe.

A life cycle analysis of a vacuum cleaner over an eight-year period (its average lifespan) shows that the majority of the environmental impacts are associated with its use – in other words, the amount and type of energy it requires.

VACUUM CLEANER

Floor coverings were once hung outdoors and beaten to remove dust and dirt, while the lowly broom did the sweeping up on floors. Now, vacuum cleaners are designed to clean all kinds of floors, with special attachments for every task.

Mechanical carpet sweepers were in use in the 1860s, and the first vacuum cleaner – Puffing Billy – was patented in the UK in 1901 by Hubert Cecil Booth. However, it was the size of a horse-drawn coach and not so convenient to use. Ten years later, janitor and part-time inventor James Murray Spangler built a suction-based creation with an electric motor (pinched from a sewing machine) to collect dirt. He showed his cousin Susan Hoover about it, and she passed the idea on to her husband, William Hoover, who went on to mass produce it. Most homes in the developed world own a vacuum cleaner.

Vacuum cleaners have become more powerful over time, increasing from motors using 500 watts (W) of power in the 1960s to over 2500W in the 2010s as consumers were persuaded that more power cleans more effectively. But it turns out that more power does not necessarily equal better cleaning, but it does result in a lower energy efficiency. To stem the ever-decreasing energy efficiency of new models, the EU passed the Ecodesign Directive which means that, as of 2017, all vacuum cleaners must be below 900W – although cordless ones are currently exempt.

THE IMPACTS

There are over 200 million domestic vacuum cleaners in the EU and they consume 18.5TWh (terra watt hours) of electricity annually. This represents 0.6 per cent of the total EU consumption and equivalent to the annual electricity generation of five gas power plants.

So, making vacuum cleaners more efficient really matters. The introduction of the EU eco-design regulation will reduce the overall environmental impacts of vacuum cleaners by 20–57 per cent in 2020 as compared with 2013. The regulation will also mean that the contribution of vacuum cleaners to global warming is expected to drop by 44 per cent compared with 2013, saving 4.8 million metric tonnes of CO2e/year (equivalent to the annual greenhouse gas emissions of the Bahamas (in 2012)). The more efficient vacuum cleaners become, and the more renewable energy we have on the grid, the smaller the overall footprint of vacuuming.

Vacuum cleaners are made of an array of raw materials, which makes them complex to recycle and contributes to the environmental impact associated with their manufacturing. They are made of aluminium, stainless steel, brass, copper and various types of plastic. As such, they fall into the electronic waste (e-waste) category meaning they are covered by WEEE regulations for recovery and recycling.

In addition, bagged vacuum cleaners create more waste than bagless models. The bags cannot be recycled, so the whole lot is bound for landfill. What's more, the packaging a new machine arrives in matters. It normally consists of a cardboard box, two cardboard trays, one interior protective cardboard and several plastic (polyethylene) bags (not currently recyclable) to protect the accessories.

Vacuuming can be bad for your health! Using a vacuum cleaner contributes in varying degrees to indoor air pollution. It has been found that the worst vacuum cleaners leave behind twice as much dust as the best-performing cleaners. A 2012 study highlighted vacuum emissions as a source of indoor exposure to airborne particles and bacteria and found that exposure rates varied widely with the make and model of the vacuum cleaner. This is why some vacuum cleaners market themselves as hygienic with super-effective filters.

WHAT YOU CAN DO

- Switch off the machine if you have to do something else briefly – most of the impact comes from the energy it uses.
- Consider using a broom if you don't have carpets, to minimise your electricity use.
- Choose effectiveness over power when buying a new vacuum cleaner. All new vacuums in the EU, bar cordless ones, have to be less than 900W. Choose a bagless vacuum cleaner to cut down on waste and save you the expense and hassle of buying bags and check it has a good filter.
- Put full vacuum bags in the rubbish bin; they can't be recycled.
- Compost the dirt from your bagless vacuum cleaner if your carpet or flooring is made of natural materials and you are only vacuuming up dust, hair and bits of food. If your carpet is synthetic, the vacuum dust will need to go in the rubbish bin.
- Switch to a renewable energy provider so that your vacuuming is emissions free.
- Make sure to take your old vacuum cleaner to a WEEE collection point for recycling.

Using disinfectant cleaning products once a week could increase the risk of developing chronic obstructive pulmonary diseases such as emphysema and chronic bronchitis by up to 32 per cent, according to a thirty-year study of nurses by Harvard University and the French National Institute of Health and Medical Research.

CLEANING PRODUCTS

Only a few decades ago the range of cleaning products was much, much smaller. Bleach was the most commonly used product and many of the specialised products we see today were not yet invented. The drive for convenience and to make dirty jobs less hassle has led to a wide range of disposable cleaning solutions, such as floor wipes and even disposable foaming toilet brush heads!

We're now discovering the effects that cleaning products are having on the environment and on our health. But, did you know you can clean almost anything with just bicarbonate of soda and vinegar (see page 68)?

THE IMPACTS

When you look at the cleaning products stored under your sink, most of them carry warning labels – they are poisonous, a skin irritant, toxic if inhaled or ingested, corrosive or hazardous to the environment. The symbol used for the 'hazardous to the environment' warning shows a dead tree and a dead fish – so it is pretty explicit.

Let's look at just a few examples:

Bleach

This diluted solution of sodium hypochlorite is a very effective killer of bacteria, viruses and fungi. You should not get it on your skin or in your eyes as it is an irritant nor use it in poorly ventilated spaces as it is not good for your lungs, especially if you have asthma.

If you mix bleach with any chemical that contains ammonia (some toilet cleaners) or acidic cleaners such as vinegar, it releases toxic chlorine gas – that's the same gas used as a weapon in World War I.

Bleach is toxic to aquatic organisms when concentrated, but when we wash it down the sink it dilutes and breaks down so that by the time it reaches rivers and lakes it has, in most cases, lost its toxicity.

Multipurpose cream cleaners

These products, designed to be used for various jobs, are made from a mix of abrasives such as calcium carbonate (as in rocks) and cleaning agents such as sodium carbonate (aka washing soda) and sodium dodecylbenzenesulphonate, a skin and lung irritant. It takes two weeks or more to break down in the environment.

Drain cleaner

All caustic drain cleaners are labelled as corrosive, flammable and harmful. They are alkaline and their ingredients, such as lye or caustic soda (also called potassium hydroxide), burn and corrode the skin on contact. When poured down the sink, the mix of chemicals turns the grease blocking your drain into a dissolvable soap-like substance.

Some drain cleaners are made using bleach, peroxides and nitrates. They are all toxic if swallowed or if they come into contact with eyes or the skin, plus they release fumes that should not be inhaled. In fact, they are so corrosive they can even damage pipes! If you have a septic tank, drain cleaners are bad news as they kill all the good bacteria working to break down your sewage and effectively stop the septic tank from working. And if they do that to bacteria, imagine the negative effect they have on aquatic life if they enter rivers, lakes and the sea.

Oven cleaner

This stuff is plain scary! It contains sodium hydroxide (a cleaning agent), isobutane (the propellant in the aerosol, which may be linked to cancer) and acrylic copolymer (a plastic to thicken it and help it stick).

Sodium hydroxide is caustic and contains corrosive ingredients similar to that in some drain cleaners, so it's a real nasty. Oven cleaner can also contain the solvent methylene chloride, which is a suspected carcinogen. The labels on oven cleaner should be enough to warn us that they are not good for us or the planet – but the ease of cleaning and convenience they offer can make them tempting to use.

Furniture polish

While the smell of furniture polish is loved by some, it hides another mix of potential nasties – methylchloroisothiazolinone and methylisothiazolinone (both of which are toxic to aquatic life), dimethicone (which can linger in the environment before it breaks down) and kerosene (poisonous if swallowed).

Window cleaner

Spray-on window cleaners often contain ammonium hydroxide, a skin irritant; it's not good news if you get it in your eyes as it burns and can even cause blindness. Breathing the stuff in is not good for you either – it can cause irritation and sneezing.

Toilet cleaner

Yet more corrosive ingredients clean the loo. Potassium hydroxide (another name for lye and caustic potash) is a drain-cleaning ingredient and is neither good for humans nor good for the environment. Whether the toilet cleaner you use is a hook-on bowl freshener or a liquid you squirt into the bowl and under the rim, when you flush you wash the mix of chemicals out into the environment.

WHAT YOU CAN DO

- Read the labels on cleaning products before you buy them or decide to use them. If they have warnings about being toxic, harmful or poisonous, I would give them a miss.
- Avoid the temptation to buy lots of specialised cleaning products. One or two cleaning products should suffice for nearly every job.
- Buy eco-friendly cleaning products from reputable brands. They may be a little more expensive but you don't need many of them – a spray cleaner, for example, can be used in the kitchen, bathroom, for dusting and even cleaning the inside of the car.
- Visit refill shops or market stalls to get the best price on eco-cleaners and reuse the same container over and over again.
- Try mechanical solutions and elbow grease instead of chemicals – a plunger, a wire coat hanger or a drain snake to unblock drains. A drain snake is inexpensive and reusable and winds down the drain – I have to say it gives great satisfaction when extracting a ball of hair and gunk from your plug hole!
- Recycle empty plastic containers after rinsing them out. Pump action spray cleaners are also recyclable. At one time we were asked to separate them out because the sprayer contained a metal spring but they can now be recycled if empty and clean.
- Try out simple DIY cleaning solutions – from lemons to remove limescale from kettles and steam cleaning ovens to cleaning windows with a mixture of half water and half vinegar. You'll soon discover the need for so many products disappears. There are whole sites devoted to such endeavours online.

HERO CLEANERS – VINEGAR AND BICARBONATE OF SODA

Bicarbonate of soda and vinegar are two hero kitchen staples with multiple uses. They do things the most advanced and specialised cleaning products can't do – and much more.

As I write this, 'someone' has managed to get slime stuck all over my lovely Donegal tweed cushion cover (it was nobody's fault apparently) – the only thing that got the slime out without ruining the cushion was vinegar rubbed gently into the fabric.

When I say vinegar I mean a very basic white vinegar – nothing fancy. The magic power comes from its acetic acid.

You can buy bicarbonate of soda (aka bread soda or baking soda) in the baking aisle of most supermarkets. It is made from sodium carbonate (mined from an ore called trona) or produced chemically by passing carbon dioxide and ammonia through a concentrated solution of salt.

Together vinegar and bicarbonate of soda react, giving off carbon dioxide that causes the mixture to bubble and foam. This combined effect offers powerful cleaning power. Try it out – it is impressive.

Some of their many uses are listed opposite – and look out for other mentions throughout the book.

TYPE OF USE	VINEGAR	BICARBONATE OF SODA
Medical	A small amount on a clean cloth cleans and disinfects cuts. Helps prevent ear infections. Dab on cold sores to reduce swelling and pain.	Mix ¼ teaspoon with water to relieve heartburn and indigestion. Add 2–4 tablespoons to a cool bath to relieve the itch of chickenpox or sunburn.
Beauty	Soak nails in vinegar before painting them for a long-lasting manicure. Dab vinegar on a pimple to help dry it out.	Dab on toothbrush to whiten teeth. Mix 2 tablespoons with water to make facial scrub. Dissolve 1 teaspoon in water and gargle for fresh breath.
Cleaning	Clean fireplaces and hearths with equal parts vinegar and water in a spray bottle. Mix 1 part vinegar and 2 parts linseed oil in a spray bottle to clean and condition leather furniture. Use a vinegar and water spray (as above) to make chrome and stainless steel sparkle.	Put a cupful in the fridge to absorb food smells. Clean and deodorise chopping board by sprinkling and scrubbing. Remove stains from cups by scrubbing with bicarbonate of soda or soaking overnight with bicarb and water inside.
Pets	Use equal parts vinegar and water to clean dog and cat beds. Mix equal parts vinegar and water and spray on your pet's coat to repel fleas and ticks. Clean dirty ears with vinegar on a cloth or tissue.	Mix 1 tablespoon into a paste with water to treat wasp and bee stings. Brush in as a dry shampoo to keep down smells between baths. Clean pet toys with 1 tablespoon of bicarb in 1 litre of water.
Garden	Mix 2 cups of vinegar with 2 tablespoons of olive oil in a bucket of hot water to clean wood panelling, outdoor furniture and fencing. Pour or spray neat on weeds to kill them off.	Treat fungal disease and greenfly on tomato plants and indoor plants by mixing 1 teaspoon of bicarbonate of soda with a few drops of vegetable oil (the oil helps the mixture to stick to the plants) in 1 litre of water and spray the solution on to the infected plants.

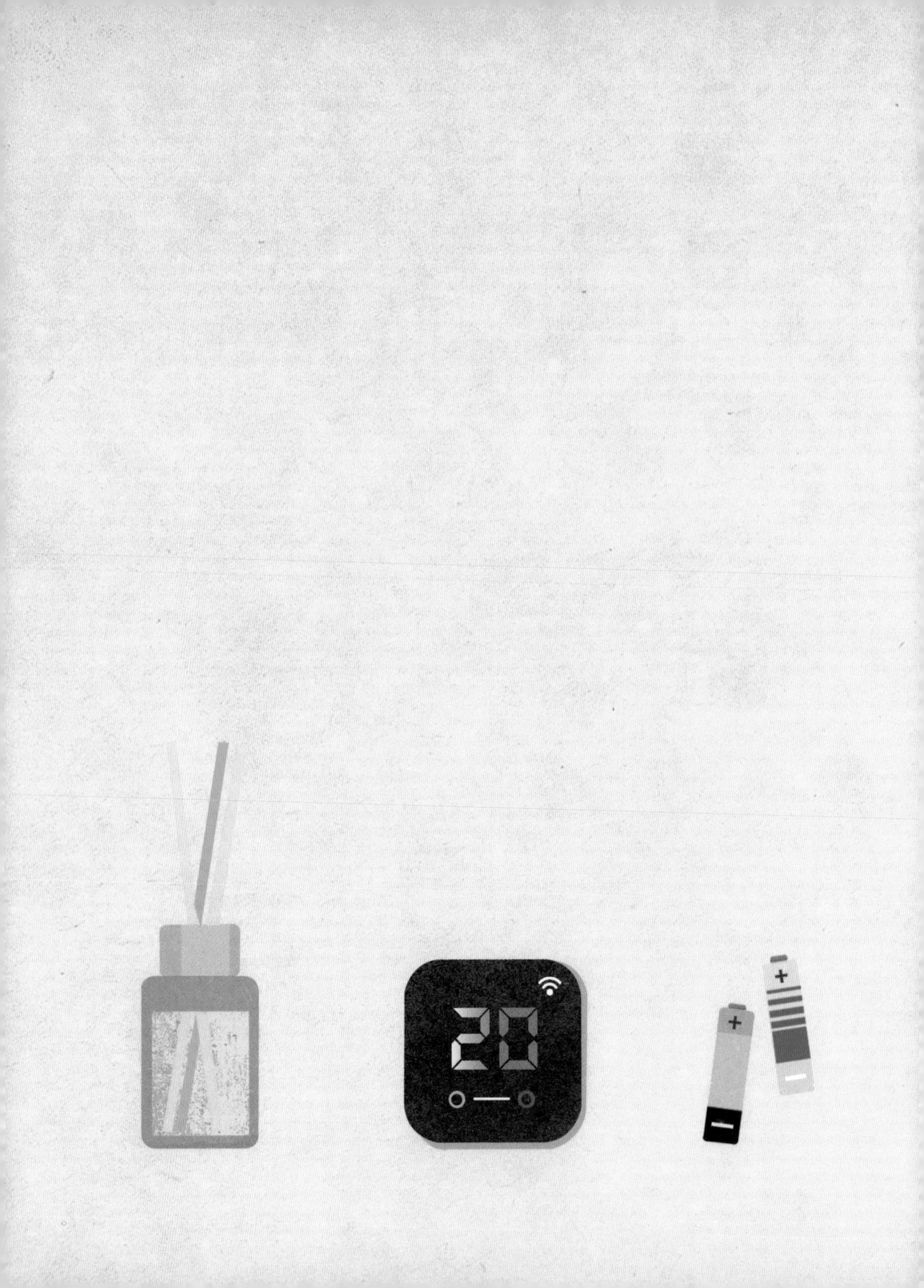
20

THE LIVING ROOM

In 2016, the world's population discarded 49 million tonnes of e-waste, including TVs. It's estimated that by 2021, that number will grow to more than 60 million tonnes a year.

TELEVISION

Televisions are ubiquitous in homes around the world, even though 100 years ago the modern appliance we are all familiar with today hadn't even been invented. It is estimated that in 2018 there were 7.6 billion connected televisions in the world – about 84 per cent of homes owned a TV.

The world's first electronic television was created around 1927, a big step up from the mechanical TVs invented in the 1800s that worked by mechanically scanning images one at a time and transmitting them onto a screen.

Different nationalities lay claim to the inventor of the TV. In the UK, John Logie Baird, a Scottish scientist, engineer and innovator, is credited with inventing the world's first television and the first publicly demonstrated colour television system.

Whatever the exact date and inventor, TVs have transformed our leisure time and access to news and entertainment. What's more, they'll continue to evolve – TVs are all likely to become smart TVs within the next ten years (electricity and internet connection permitting), so everyone can watch what they want, when they want and even via virtual reality – by wearing a set of smart glasses.

Over time the relative cost of owning a TV has fallen dramatically – by more than 96 per cent in the last seventy years. Almost all households in the developed world own a TV, while ownership is lower in developing countries and particularly in Africa where fewer than one-third of households own a TV set due to affordability and access to electricity.

Modern TVs can have plasma or LCD screens. Both use phosphor-gas-filled cells. In plasma screens, the gas is sandwiched in its plasma form between two glass layers along with electrodes. LCD TV screens also use two glass layers, where one is coated in liquid crystal. In both cases the main component is glass, although plastic can be used instead of glass in smaller TVs. The electronics and case are made of various materials – plastics, copper, tin, zinc, silicon, gold and chromium.

THE IMPACTS

Manufacturers of LCD flatscreen TVs use a potent greenhouse gas called nitrogen trifluoride (or NF_3), which is 17,000 times more potent in its warming potential than carbon dioxide and persists in the atmosphere – for up to 740 years. Even though NF_3 is used in tiny quantities in such TVs, the production and industrial use of NF_3 has soared since 1990 – in fact, there was a 1,057 per cent increase in US annual emissions of NF_3 from 1990 to 2015, according to the Environmental Protection Agency, which is related to the manufacturing of TVs.

What's more, there are concerns that more of the NF_3 gas is leaking into the atmosphere than was previously thought.

TVs are made of many components such as metal, heavy metals (mercury, lead and cadmium), glass, plastic, solder, silicon and fluorescent chemical coating. Their production uses energy and natural resources, all of which contribute to the footprint of the product. And that's before it is packaged, transported and plugged in to use electricity throughout its lifetime. At the end of its life a TV continues to have an impact – as electronic waste (or e-waste) it is hazardous and needs to be recycled carefully.

Until 2018, China accepted 70 per cent of the world's e-waste. After 2018, Vietnam and Thailand soon became important destinations until their ports were overwhelmed and they soon followed suit, limiting their intake of e-waste. Increasingly countries need to take

TV VIEWING FIGURES

Here we show how much TV people in different countries watch per day. Figures are minutes per day spent watching TV.

responsibility for their own e-waste and this requires putting systems in place to separate the electronics into their component parts for recycling. In fact, it's cheaper to reclaim metals from TVs compared with mining the planet's virgin resources. Researchers in China found that mining copper, gold and aluminium from ore costs thirteen times more than recovering the metals via the 'urban mining' of e-waste.

It is estimated that 569,000 tonnes of LCD TV waste has been produced globally up to 2018 and that most of this e-waste ends up in landfills, illegal dumps or being burned (burning is used to melt away plastic and expose valuable copper wire in illegal dumps). As a result, pollutants are released into groundwater, the soil and the air with knock-on effects on people's health and that of the wildlife.

TVs consume energy when they are turned on and also when they are in standby mode (see also page 89), which costs you money and contributes to greenhouse gas emissions (unless you have a renewable energy supplier).

WHAT YOU CAN DO

- Choose a small TV – a smaller TV uses fewer raw materials and requires less energy to run. The bigger your TV, the bigger your energy bill. Plasma TVs use more energy; LCDs are better. For example, an energy-efficient 32-inch LCD TV will typically use half the power of a model with a 42-inch plasma screen.
- Look for the energy label on TVs to see how they compare in terms of energy efficiency. The scale runs from G, the least efficient to A+++, the most efficient.
- Repair your TV if it develops a fault. Look for a local professional or repair café and give your TV a longer life.

THE FUTURE OF TELEVISION?

The TV of the future may be curved, not flat, or integrated into furniture, and it will likely talk to us and help us decide what to watch. In fact, we may not have TVs at all! Instead electric devices might simply beam images onto walls and ceilings, bypassing the need for a screen altogether.

The question 'What's on the box?' or 'What's on TV?' is no longer valid. Most people now choose what they want to watch when they want to watch it. In 2019 Netflix alone had more than 139 million subscribers. That's a lot of content to store in data centres that are a growing consumer of electricity (see also pages 88–89 and 98–99).

- Recycle your old television. In developed countries there are 'take back' systems for old TVs operated by some manufacturers or e-waste programmes such as WEEE in Europe (see also page 96, and opposite) that recycle old sets and prevent them from going to landfill.
- Switch off your TV – don't just put it on standby! – when no one's watching it (see also page 89). Consider plugging your TV, DVD player and network box into a power strip, so you can turn them all off with one flick of a switch.
- Recycle the packaging your TV comes in. TVs tend to be delivered in cardboard boxes with soft plastic wrap and expanded polystyrene beads or board. Cardboard can go in your household recycling bin but the expanded polystyrene has to be taken to a specialist recycling point (which are not available in all countries).

MANAGING E-WASTE ON SMALL ISLANDS

Faced with a growing electronic waste or e-waste problem on two small islands with limited landfill space, a group of local organisations in Antigua and Barbuda got together in 2013 to try to find a solution to the problem.

Concerned about the toxic effects of e-waste on their local environment, this group was motivated to reuse and refurbish electronics, especially computers, for resale and reuse. So, the Antigua Barbuda E-waste Management Center was born.

It was created as a collaboration between Computer Reset (a small business providing computer training and selling new and used computers), the Boys Training School (a state-run rehabilitation school for boys) and Gray's Farm Baptist Church.

The e-waste centre accepts communications equipment, batteries, mobile phones, computers, transformers, computer cables, photocopiers, inkjet cartridges and small household electrical appliances from individuals and businesses. They sort the items and refurbish what they can for use in the Boys Training School – the computers and training provided at the centre help the young men gain new skills and livelihood opportunities. Other refurbished computers are sold through Computer Reset's shop to fund the work of the centre.

Any item that can't be repaired or reused is taken apart and sorted into components and materials for recycling. Most of the materials are sold to facilities in Canada for recycling as the islands do not have their own specialised recycling facilities.

Through its outreach and public education campaigns, the e-waste centre has raised awareness of the problems caused by e-waste, which has encouraged local people and businesses to donate their old technology and devices for recycling instead of throwing them away.

So, it's good news all round – for the environment, for the young men and for all the businesses and individuals that can use this recycling service.

People in Ireland spend approximately 7.3 hours a day sitting down. About half of this is at work, school or college but the rest is at home, and most probably on the sofa.

A large survey in England, conducted in 2012, revealed that 30 per cent of adults spent at least six hours of every weekday sedentary, rising to 37 per cent of adults on the weekends.

SOFA

It is hard to imagine your living room without a sofa. They are our go-to place for relaxing and watching TV or spreading out with the weekend papers. Nowadays, it is common to have a sofa and a range of armchairs, bean bags, dining chairs and stools in a home. It used to be that metal and wood were the commonly used materials in furniture, but now plastic is increasingly used for both the frames and the padding in the form of foam.

The Egyptians were making sofas for the very wealthy way back in 2000BC, and the word 'sofa' is said to come from the Arabic word 'suffah' for a low bench. However, it would take thousands of years before sofas became mainstream. European furniture makers' first efforts looked like giant armchairs in the 1690s, but were only for the elite; it wasn't until the early 1900s that sofas became more available for the masses. Early sofas were pretty uncomfortable, stuffed with horse hair, straw and moss, with rigid wooden frames. Today's comfy sofas benefit from the advent of cheaper synthetic textiles and foam, but at what cost to the planet?

THE IMPACTS

Sofas are a household essential and so it is not about not using (or avoiding) sofas, but it is worth thinking about what they are made of and how we use them. Interestingly, the furniture in our homes and offices is responsible for a significant portion of the carbon emissions associated with these buildings. A 2017 study found that furniture is responsible for around 10 per cent of greenhouse gas emissions emitted by a building striving to be zero emissions. The emissions relate to the production of the furniture, for example from the trees cut down to make wood, the energy used to make the metal and the oil and energy used to make the plastic foam and synthetic upholstery.

Most sofas that get thrown out could be used again, with a little cleaning, new upholstery or repair. Yet, only 17 per cent of sofas in the UK are reused in some way. The rest are incinerated, go to landfill or are broken down into their component parts for recycling.

The foam that's used in the manufacture of sofas and chairs and the polystyrene beads inside bean bags are particularly problematic. The beads in a bean bag are made from expanded polystyrene or polypropylene, both of which are plastics so do not biodegrade. Polystyrene is recyclable but there are few facilities that accept it, so it tends to end up in landfill or being incinerated.

In the past the manufacture of the synthetic foam used in cushions and bean bags used to contribute to the hole in the ozone layer, but different chemicals are now used to inject air into the foam to reduce this problem. Burning polystyrene and foam releases dioxins, which cause hormonal and immune problems in people, so illegal burning of old furniture poses a real health risk.

Most sofas are treated with chemical fire retardants as the foam and fabric is often very flammable and are required to have a certain level of fire resistance by law. The chemicals make the sofas less flammable but one study found that toxic gases were released when it was on fire. All this makes buying a sustainable sofa a complex task.

WHAT YOU CAN DO

- Steam clean your sofa or armchair to perk it up rather than buy a new one. You can hire steam cleaners for a day or weekend, you don't need to shell out to buy one yourself; or ask neighbours, in case anyone has one already and you can borrow it.
- Reupholster your sofa to give it a second life. Learn how at a Repair café (see page 187), take an upcycling course or look locally for a specialist who can help. Choose naturally water-resistant fabrics, such as wool, tweed, cotton and linen.
- Rehome your sofa when you don't want it anymore. Sell it online or give it to a local charity or furniture reuse network.
- Buy a second-hand sofa if you are looking to upgrade yours rather than buy brand-new. Online buy-and-sell sites often have sofas that are in great condition.
- Buy a sofa made from sustainable materials. Look for retailers and manufacturers that use FSC-certified wood, recycled materials and natural textiles, such as wool and linen. Buy as locally as you can to reduce the carbon footprint linked to transporting the sofa.
- Look for bean bags filled with buckwheat hulls or other natural ingredients instead of polystyrene. Dispose of polystyrene beans carefully. Ask at your local amenity site about specialised polystyrene recycling centres.
- Never burn old furniture. The emissions released into the air are bad for you and bad for the environment.

Additional fact

Reusing a sofa by selling it on online or passing it on to a charity shop can save approximately 55kg CO2e per sofa, equivalent to driving 134 miles (216km) in an average petrol car.

TURNING PLASTIC WASTE INTO FURNITURE

Dunia Designs, based in Arusha, Tanzania, is a company that makes furniture, including sofas, from at least 90 per cent upcycled or recycled plastic material.

The company employs local people to collect plastic bottles and plastic bags, both of which are a real litter problem in Tanzania. The plastic bags are shredded to be used as filling in sofas, cushions and bean bags while the plastic bottles are shredded and processed to produce recycled plastic 'wood' that is then used in the construction of sofa frames, chairs and tables.

This gives work to ten full-time staff who also make beehives, school desks for government schools and permaculture planters from recycled plastic.

Dunia Designs have contributed to forest conservation by avoiding the use of 560 tonnes of wood in furniture making and using recycled plastic instead – this equates to 130,000 tonnes of CO_2e per year. They also invest some of their profits in protecting 15 acres of indigenous forest in Tanzania with the World Land Trust.

The company collected 560 metric tonnes of plastic waste in their first four years in operation and transformed it into furniture and a profitable business.

They sold €400,000 of furniture between 2015 and 2019, showing that it makes business sense to adopt a circular economy approach. They were even invited to the UN Environmental Assembly in 2019 as a best-practice model of responsible and ethical business.

Despite what we might think, almost all, 92–94 per cent, of the carpeting used in people's homes today is made from plastic not wool. Nearly 100 per cent of the carpet used in commercial environments is plastic.

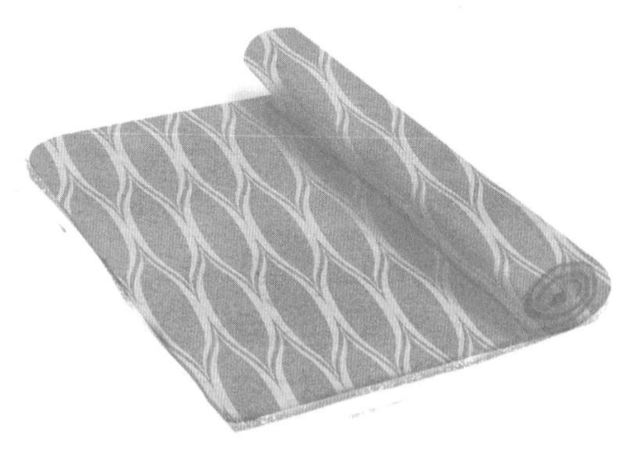

CARPET (AND OTHER FLOORING)

On average, we spend 85–90 per cent of our time indoors with some kind of floor covering under our feet. Floor coverings not only add comfort but also keep us warm or cool underfoot. Carpet is loved by people who appreciate its warmth and comfort and hated by those who are worried about the dust mites it may be harbouring. So, what you choose as a floor covering in your home is a very personal choice.

While carpets' origins lie in wool, they are increasingly made from synthetic materials, which are cheaper and more durable. Wood is also used as a floor covering as is natural stone, ceramic tiles, plastic laminate, vinyl and linoleum. Each material has its pros and cons from a functionality point of view and also in terms of environmental impact.

THE IMPACTS

Wool carpets have a lower environmental footprint than those made from nylon; the switch to plastic has been a gradual one starting in the 1950s. Despite its name, wool carpet is made from more than just wool and often contains some polyvinylchloride (PVC), though it does contain significantly less plastic than synthetic carpet.

In 2018, a life cycle analysis of a nylon carpet versus a wool carpet revealed that manufacturing a nylon carpet uses 80 times more energy and produces 49 times the CO2e of a wool carpet. Nylon carpets also use non-renewable petroleum resources, while natural fibres – such as jute and cotton – can be grown, harvested and grown again. These natural fibres are not without their own impacts, though, as they require land, fertiliser and water to be grown and also contribute to pollution via dyeing, manufacturing and transport.

Natural wooden floors are a good choice but it's important to make sure the wood is recycled or it comes from a sustainable source and is FSC-certified. Treating and transporting wood contributes to carbon emissions, so reusing or refurbishing wooden floors is better for the environment than buying new. You can now get bamboo wooden floors that are from a renewable resource, but care will be needed as demand for bamboo grows to make sure that it does not compete for land with food crops (leading to food shortages) needed to feed the world's growing population.

Linoleum is reasonably sustainable; also called marmoleum, it is made from linseed oil, cork dust, tree resins, wood flour, pigments and ground limestone, is very hard-wearing and should last twenty-five to forty years. However, the name 'lino' is also used for vinyl floor coverings made from plastic. Cork is another natural floor covering that is hard-wearing, long-lasting and warm. It can be made from tree bark or recycled corks. Ceramic tiles are estimated to have twice the environmental impact of marble tiles due to the greenhouse gas emissions associated with their manufacture and heavy metals that may be used in glaze. The source of the marble or other stone tiles is important, as quarrying can have an negative impact on the environment.

Natural stone is a sustainable choice, especially if it is sourced near where you live and doesn't have to be transported a long distance. Natural stone has a lower carbon footprint than ceramic tiles, parquet, PVC, lino and carpet.

A 2019 study of eco-friendly floor coverings ranked flooring choices. First is flooring made from plant-based renewable materials, such as wood, cork or linoleum. The plants store carbon while they are growing and don't need a lot of processing to be made into flooring.

In second place came vinyl made from recycled plastic or tiles made from recycled glass. And in third position, and bottom of the list, is carpet – synthetic or wool. The methane emissions associated with sheep rearing and wool production are a significant contributor to climate change and increase the carbon footprint of a wool carpet.

A SUSTAINABLE RUG

What if waste fabric and offcuts could be turned into beautiful, designer rugs? This was the challenge Katarina Brieditis and Katarina Evans took on and which became the Re Rag Rug project: to produce twelve rugs using twelve different techniques in just one year. They took discarded waste and otherwise worthless material from countries facing a growing textile waste problem and experimented with a wide range of rug-making techniques that don't require lots of machinery and that could be replicated by small businesses in the countries the waste originated in. The twelve rugs are part of an exhibition to demonstrate how design can lead to socially and environmentally socially sustainable products. The rugs can be used on floors, as a cosy wrap or hung on walls to reduce draughts and absorb noise. The next step the Swedish design duo face is to find ways to scale up production with local craftspeople and to bring the rugs to market with retail partners.

Disposing of old carpet and other flooring in an environmentally neutral way also presents a challenge. A huge amount – 1.6 million tonnes – of used carpet is thrown away in Europe every year, going to landfills and for incineration. Almost 600,000 tonnes of flooring, mostly carpet, is disposed of each year in the UK, and only 2 per cent is recycled. According to waste and sustainability charity WRAP UK a small quantity is incinerated but over 90 per cent goes to landfill. Carpets take up lots of space in an ever-decreasing number of landfill sites and produce climate changes gases. As the carpets break down, they release toxins into the landfill leachate that need to be carefully managed to avoid pollution.

Carpet also contributes to indoor air pollution. That new carpet smell is due to low levels of the chemical 4-phenyl cyclohexane, which luckily is non-toxic but is released in the first few weeks of a carpet's life. And research shows that carpets harbour allergens, including housedust mites and fungi.

WHAT YOU CAN DO

- Choose natural floor coverings over synthetic ones: opt for wood, cork and stone over wool. Alternatives to wool carpets are those made from sustainably harvested sisal and sea grass. New innovations mean that up to 3,000 plastic bottles can be recycled and given a new life as a rug.
- Consider alternatives: substitute plastic laminate with reclaimed or recycled wood and opt for marmoleum or cork instead of vinyl.
- Maintain and refurbish wooden floors instead of replacing them. Sand them down to bring them back to their best.
- Consider tiles made from natural stone as an alternative to ceramic tiles.
- Look for fairtrade rag rugs made from discarded fabrics.
- Dispose of old carpets carefully. If you can, reuse old carpet – to suppress weeds or to insulate a compost heap – or find a new home for it, such as at a local animal shelter. Contact the Reuse Network online and see if there is someone near you who will take your old carpet and put it to another use.

Additional facts

The UK flooring sector is estimated to produce 290 million square metres of material annually, equivalent to covering 85 full-size football pitches every day.

A 2017 study conducted in France estimates that recycling 100 square metres of carpet prevents 445kg CO2 from being emitted into the atmosphere (the equivalent of driving over 990 miles (1593km) in a car), diverts carpet from landfill, reduces the need for new raw materials and creates green jobs. There is a general lack of carpet recycling facilities in the UK and Ireland, but other countries, such as Canada, Japan and the United States, can already recover nylon from used synthetic carpet but not yet from wool carpet.

An open fireplace emits eight times more climate-change-causing carbon dioxide than an efficient wood-burning stove and most of the heat escapes straight up the chimney.

There are more than 1.5 million wood-burning stoves in the UK. It is estimated that one wood-burner generates more particulate matter per hour than eighteen diesel passenger cars.

FIREPLACE

A fire or wood-burning stove brings warmth on a cold night but its effect on us is more than just heat. A open fire, in particular, has a distinctive smell, it crackles and snaps as it burns, and all fires, whether contained within a wood-burner or not, are hypnotic to watch. The combined effect creates an emotive experience that makes us nostalgic and gravitate towards a fire.

Since early humans discovered fire in the Stone Age we have been drawn to it for safety, for heat and for cooking. When fire pits moved indoors the need for ventilation to let the smoke out arose and led to the introduction of chimneys and fireplaces.

While modern new-builds, such as passive and A-rated eco-houses, no longer have fireplaces and use zero-emission heat sources, older houses (especially those in cooler countries like Ireland, the UK and New Zealand) tend to have open fireplaces or wood-burning stoves. In cities in the UK, for instance, most fires are burned at the weekends, indicating they are almost certainly for decorative purposes.

THE IMPACTS

Both open fires and wood-burning stoves contribute to climate change and to air pollution. Burning coal and peat is particularly bad from a climate change perspective as both contain carbon that is released into the air on burning (see also pages 88–89). While wood is less damaging in terms of climate change, especially if new trees are planted to replace those cut down, it is still a major contributor to air pollution.

Burning any type of fuel produces soot or what's known as 'black carbon' and particulate matter (PM) – these fine particles contribute to global warming and air pollution. In the atmosphere black carbon particles absorb energy and cause warming and when black carbon combines with snow and falls to the ground, the snow absorbs sunlight rather than reflecting it and this further contributes to warming.

The microscopic particles known as PM2.5s, which can come from burning fuel, can enter your lungs where they affect your breathing and have been linked to asthma and other health conditions. Diesel cars, power plants and wild fires are other sources of PM2.5s.

The health effects of air pollution are significant. The NHS in England estimates that almost 30 per cent of preventable deaths in England are due to diseases specifically attributed to air pollution. A staggering 28,000–36,000 premature deaths in the UK every year are linked to polluted air. In Ireland 1,180 people die prematurely from air pollution every year.

So, open fires and wood-burning stoves can be bad for your health. The smoke and fumes they give off contain everything from formaldehyde and benzene to soot particles that lodge in the lungs. Old wood-burning stoves give off up to fifteen times more toxic smoke and four times as much CO_2 as modern, more-efficient models depending on the fuel burned. Poor ventilation compounds the health risks, including from carbon monoxide poisoning. There is no question about the carbon monoxide facts – that's why we are all encouraged to have a carbon monoxide alarm if we have a stove or an open fire.

WHAT YOU CAN DO

- Don't burn coal or peat in your fire or multi-fuel stove; they are the most damaging from an environmental and a health perspective. A good-quality wood-burning stove is better in terms of climate change and heat efficiency than an open fire burning coal, wood or peat, but it still produces air pollutants.
- Burn only dry wood. Dry wood produces less air pollution than wet wood.
- Source your wood locally so that the carbon footprint of transporting it is reduced. Do check that the wood comes from a sustainably managed forest.
- Clean your chimney regularly to make sure your wood-burning stove is working as efficiently as possible.
- Light your fire less if you really only need it for comfort rather than heat.
- Apply for a grant to retrofit your house and install a heat pump instead of a fire or wood-burning stove. That way, you can reduce your emissions to zero.
- Seal up any unused fireplaces so that you are not losing heat up the chimney.

Additional facts

A 2017 study found that 23–31 per cent of PM2.5s in the urban conurbations of London and Birmingham was due to wood burning.

The World Health Organization estimates that 7 million people worldwide die every year from exposure to fine particles in polluted air which come from burning fuel, traffic and industry.

Two million people in London are living with illegal levels of air pollution, primarily due to vehicle emissions, but also due to domestic fires and stoves.

In 2016 a nationally representative survey of the US population found that 20.4 per cent of the population reported health problems from exposure to air fresheners and deodorisers.

Fewer than 10 per cent of the ingredients in air fresheners are typically disclosed to consumers.

AIR FRESHENER

Nowadays, air fresheners come in various forms – as plug-ins that claim to 'purify' air, paper 'trees' that make your car smell like a pine forest and deodorising sprays to neutralise bad smells around the home.

Freshening the air is not a new idea; ancient civilisations – the Egyptians and Chinese, to name just two – used incense, flowers, fragrant woods, spices, herbs and citrus fruit to disguise bad smells and even to try to eliminate disease.

The first commercial air freshener designed for use in homes, Air Wick, was sold in the US in 1943, and its use quickly spread to Europe, Canada and Australia. In 1956, Glade, the first air freshener in an aerosol can, hit the supermarket shelves and is still on sale today.

In Europe aerosol air fresheners are used at least once a week in the homes of 39 per cent of people, while 40 per cent of people use plug-ins and 30 per cent use passive air fresheners (the liquid type, for example). These figures already seem high but they rise again when survey respondents were asked about monthly rather than weekly use of air fresheners: 89–94 per cent of people in Europe use air fresheners at least once a month. That leaves very few people living air-freshener-free and when you consider that, on average, people spend about 90 per cent of their time indoors that's a lot of time breathing in air-freshened air.

THE IMPACTS

Air fresheners require raw materials and lots of energy to be produced and packaged. Spray aerosols use aluminium and propellant gases, and plug-in and liquid air fresheners' containers are made from plastic, which consumes petroleum and does not biodegrade. While empty aerosols can be recycled with metals, most plastic air fresheners are not recyclable due to the mix of plastics and any residue of the fragrance deems them not clean. Since plug-in air fresheners have a plug, they should only be recycled via WEEE or e-waste recycling. The glass or metal holders of scented candles are recyclable but only if you can get all of the wax out.

When it comes to what's inside an air freshener, it's difficult to know as there is no requirement for the manufacturer to list the ingredients on the packaging. Some air fresheners emit over 100 different chemicals, including volatile organic compounds (VOCs; see also page 170) and phthalates, which have been linked to various health problems (see page 26). Some air fresheners have also been found to contain chemicals such as acetaldehyde, which is thought to be a human carcinogen and 'hazardous air pollutant'. The accumulation of these substances in the air contributes to indoor and outdoor air pollution – and air pollution is the cause of 7 million premature deaths each year according to the World Health Organization.

Over the course of the year, an ordinary plug-in air freshener will use about 18.4kWh of electricity, which is more energy than a powerful cooker extractor hood used for 40 minutes per day for a year would consume.

Unfortunately, naturally and organically sourced air fresheners are not always the answer. Patchouli oil, reed diffusers, incense and scented candles, for instance, all contain VOCs. It's best to use any of these air fresheners in well-ventilated rooms to avoid the risk of indoor air pollution.

WHAT YOU CAN DO

- Open the windows and let the fresh air in (assuming the air outside is a decent quality).
- Find alternatives. Neutralise bad odours with white vinegar or bicarbonate of soda. Put some white vinegar on a cloth and wave it around a room to dispel smells; and bicarbonate of soda in a cup absorbs smells in fridges and can be sprinkled in bins to deodorise them. Bake or cook something that smells good or grind some coffee beans.
- Make your own air fresheners: place empty vanilla pods in vodka and then dab it on to fabric to release its sweet smell; citrus peel left to dry out releases super-fresh scents; make lavender bags to freshen up drawers.
- Avoid plug-ins and plastic holders for liquids and the overpackaged and hard-to-recycle forms of air fresheners. If there are no ingredients listed, steer clear.

Additional fact

A 2016 survey in Australia found that 73.7 per cent of respondents were not aware that air fresheners, even those labelled 'green' and 'organic', can emit hazardous air pollutants. Over half of those surveyed said they would no longer use a product if they knew it could be harmful.

Surprisingly 40 per cent of us don't know how to work our programmable thermostats and instead use it as if it were an on-off manual thermostat, which means we are missing out on the benefits in terms of the convenience of the heating going on and off automatically and the potential costs savings of using heating or cooling more efficiently.

THERMOSTAT

Thermostats are the cause of many a marital, family and shared household row. Who turned it up? Who turned it down? The house is too hot or too cold and the inevitable arguments over heating bills ensue! In fact, thermostats should make our lives easier, save us money and reduce our carbon emissions – that is, if we use them properly.

Heating and cooling is a major contributor to household bills. Energy accounts for, on average, 4 per cent of total household expenditure in the UK (2016); for families on a low income, heating can be more than double at 8.4 per cent of household expenditure. This is due to several things – energy being relatively expensive, poorer-quality housing being less well insulated and low-income families having less money overall.

In Ireland, 61 per cent of household energy use in 2016 was attributable to heating and 19 per cent to water heating. So, making your heating run more efficiently will undoubtedly reduce your energy bills.

THE IMPACTS

Not having a thermostat or not using it properly costs you money and increases your carbon footprint. Every additional degree on your thermostat increases your carbon footprint by 320kg CO_2 (based on living in a three-bed semi in the UK). Fourteen per cent of UK greenhouse gas emissions come from the residential sector and heating is responsible for up to 85 per cent of domestic energy use, so heating your home is where individual actions really do count.

So, what are we getting wrong when it comes to using our thermostats? Many of us set the temperature too high – the recommended winter setting is between 18 and 21°C. Also, where is your thermostat located? If it is in a warm room – the kitchen or a sun trap – it can shut off before the rest of your home is warm. Or, conversely, if it is in a draughty hall, all of the other rooms will feel tropical as the boiler struggles to heat the hallway.

Your house should be cooler when it's empty – so having the heat always on is not more efficient. Hotter houses have more heat to lose to the outside, so you will just waste more energy. What is important is to get the timing right, so that the heat is on when and where you need it. If you have radiator thermostats, set them to a level you find comfortable in the main living space (which should be mid-way on the dial on the radiator valve) and make sure the radiators in other rooms are set to a lower level, as you aren't using them as much; bedrooms should be cooler for sleeping.

JUST SET THE PROGRAMME

The first thermostat-type heating control was invented way back in 1885 by Albert Butz to control the heat from coal furnaces. His patent was bought in 1906 by another inventor, Mark Honeywell, who made the first programmable thermostat called the Jewell. It allowed users to set a clock to turn the heat down at night and come on automatically in the morning. By 1953 the modern programmable thermostat with a round dial, called a Honeywell, was launched to control heating in homes.

WHAT YOU CAN DO

- Turn down your thermostat so that it sits between 18 and 21°C. If you get a little cold, pull on a jumper rather than turning up the thermostat. Keep a blanket near the sofa for when you feel chilly. Remember when you turn down the dial by 1°C you save money and reduce your emissions.
- Learn how to programme your thermostat; look up the manual (you should be able to find it online if you've lost the copy that came with it when you bought it) or watch a YouTube video (I find this easier than poring over a manual) or ask someone who knows to show you how to use it. Modern thermostats allow you to heat different parts of the house at different times and to different temperatures and to have different settings for weekdays and weekends.
- Install a thermostat in your home – either on the radiators or a wall-mounted room thermometer. Get the thermostat that suits you. If you can only manage an on-off manual system and that works for you, that's OK. If not, get a programmable thermostat so that your home is warm only when it needs to be.
- Install a smart thermostat that allows you to connect to your smartphone and control your heating remotely. Some systems are self-learning and programme themselves based on observations of your habits and schedule, so it does all the thinking for you.

Additional fact

The Energy Saving Trust in the UK estimates that turning the thermostats down by 1°C in a typical three-bed semi can save £80 per year, and you can save the same amount again by installing room or radiator thermostats and using them properly. If you live in a warmer climate, the same rules apply for cooling your house; adjusting the thermostat – even just 1°C – makes a big difference.

ENERGY AND ELECTRICITY

Every time we turn on the lights, plug in the kettle and charge a mobile phone we use electricity. It is only when we have a power cut or our phone battery dies that we realise how fundamental our reliance on it is.

The fact that we use the internet for ever more aspects of life means that the electricity demand associated with digital and data is soaring. The scope of what a digital life entails is ever expanding – we use the internet to listen to music, to communicate, to watch TV, to read, to store photos, to do our banking, shopping, research and work.

Globally, data centres consumed approximately 3 per cent of energy and accounted for 2 per cent of global emissions in 2015 – this is expected to triple by 2025. Add to that the fact that more people around the world are gaining access to electricity (which they absolutely need) and the fact that heat and transport will be electrified as we move to a zero-carbon future and it is clear that demand for electricity will only grow.

WHAT ARE RENEWABLES?

Renewable energy is made from natural resources that can be replenished – the sun, the wind, tides, waves, water and geothermal. They produce electricity without CO_2 emissions and air pollution. Bioenergy uses plants and agricultural waste and to create energy and emits CO_2 in the process, so new plants have to be planted to absorb the CO_2 to achieve carbon neutrality.

Currently, most electricity comes from burning fossil fuels (coal, oil, gas and peat), all of which release climate-changing CO_2. In 2018 coal was still the main fuel being used to create electricity around the world, followed by natural gas and then oil, with renewable energy making up 25 per cent of total production.

Whether renewable energy comes via solar panels, wind turbines or the ground, we need more of it. According to the International Energy Agency, renewables should produce two-thirds of the electricity required globally by 2040, with a mirrored sharp decline in the use of coal and oil in particular. At the same time, energy efficiency has to improve, so that we both use less and waste less energy.

READY TO DIVEST?

When you choose a renewable energy provider you are investing your money in zero-emissions energy instead of fossil fuels. This sends a signal to the market that customers want more clean energy and makes less finance available to fossil-fuel companies. Similarly, divesting from fossil fuels means making sure that none of your money or your investments (such as in a pension fund) support the fossil-fuel industry. Request an ethical or responsible pension instead. If investing money look for green investment options and make your money work for a safe future.

BE MORE ENERGY EFFICIENT

While governments are in the driving seat when it comes to creating energy policy, in the UK 40 per cent of national emissions come from households – so changing what you do at home matters.

- Switch off standby. Appliances left on standby still use power – so-called vampire energy. Globally, standby power is responsible for 1 per cent of global CO_2 emissions and in European countries vampire electricity is responsible for about 11 per cent of the total annual electricity consumption per household. So, that's 11 per cent of your electricity bill you could be saving.
- Insulate your home to reduce heat loss. There are grants you can apply for to cut the cost of insulating walls and attics and to install double-glazed windows. If you cannot afford to carry out those kinds of modifications, close your doors and pull your curtains at night to reduce draughts and prevent warm air escaping. Seal up unused fireplaces.
- Turn down the thermostat. You can reduce your heating bill by 10 per cent by turning down the heat by just 1°C. If the water in your taps is so hot you have to add lots of cold water, turn down the thermostat on your boiler, as well.
- Read up on your appliances. Dig out the manual that came with it and find out how to select eco mode as the default. Run dishwashers and washing machines only when they are full and select an eco-mode and low temperature.
- Make sure your fridge isn't too cold; it should be set between 2 and 3°C to preserve food without freezing it. Your freezer should be set to −15°C.
- Don't overfill your fridge or it will struggle to work efficiently, requiring more electricity, and never put hot or warm food in the fridge or freezer as it will have to work harder and consume more electricity to regulate the temperature. Likewise, keep your fridge and freezer defrosted, and don't leave the fridge door open.
- Switch off and unplug appliances at night and when away from home. Turn off your computer at night or if you are not using it for an hour or more. Unplug things that are not in use – hairdryers and hair straighteners, mobile phone and tablet chargers, coffee machines and toasters, printers and scanners.
- Turn off the lights when leaving rooms and make sure all your lights are off before you leave the house.
- Service your boiler regularly. It will save you money and work more efficiently, producing fewer emissions.
- Use a power strip so that you can turn off multiple appliances at the same time.
- Install smart heating controls that allow you to set the temperature and the time the heat goes on and off and to control the heat in your home from your smart phone.
- Switch to a day and night electricity meter if you use appliances like your washing machine at night and have an EV to charge. This will allow you to access lower-cost night-time rates.
- Switch to a renewable electricity supplier.
- Consider installing solar panels on your roof. Look for grants and favourable bank loans to help with the cost.
- Keep the lid on. When boiling food in a pan keep the lid on to use less energy. Boiling a pan of potatoes with the lid on uses almost half the energy of boiling them with the lid off.
- Choose the right-sized pan and ring when cooking. Use the smallest pan and ring you need for the job.
- Make sure your oven door seals well and keep your oven clean so that it performs to its best.
- Use a smaller top oven if you have one and are only cooking one thing.
- Make the most out of turning on your oven and batch cook several things.

Incredibly, the average home has 193 electrical appliances and 110 batteries.

BATTERY

Batteries power everything from watches, TV remote controls, mobile phones, hearing aids, cars and toothbrushes. Making sure said batteries are charged can feel almost like a full-time job sometimes!

Single-use batteries are convenient, cheap and last longer than rechargeable batteries, but they are ultimately disposable and potentially toxic. If you leave batteries in a device that you don't use very often, you may notice they leak a corrosive fluid.

There are three types of single-use battery: zinc-carbon batteries are the cheapest batteries you can buy but they don't last very long; alkaline batteries have more energy and last longer (they can hold their charge if unused for many years, so they can be super-useful in an emergency); and button batteries are the little, round ones used in watches and hearing aids and work much like alkaline batteries do.

Rechargeable batteries can be used and recharged again and again, and include the lead-acid batteries used in vehicles and lithium-ion batteries used for portable electronics such as cameras and smartphones. You can easily buy rechargeable AA and AAA batteries and charging units from many hardware or department stores.

THE IMPACTS

Batteries contain a cocktail of chemicals including a mix of heavy metals, such as mercury, lead, cadmium and nickel. Left in the environment these metals can leach into the soil and groundwater causing pollution. If sent to landfill or for incineration, their heavy metal content mean that batteries are considered hazardous waste and have to be carefully managed so that they don't cause harm.

All batteries need to be disposed of carefully

to prevent toxins escaping into the environment.

Recycling schemes are being established around the world to encourage people and companies to recycle old batteries. According to the European Commission, approximately 160,000 tonnes of consumer batteries (single-use and rechargeable) enter the European Union alone every year; 46 per cent were collected for recycling.

Initiatives in some US states, South Africa and Australia among others have been set up to encourage battery recycling but in large parts of the world with no recycling facilities or even managed landfills, old batteries simply end up in the environment.

The EU introduced legislation in 2005 to allow consumers to recycle their batteries free of charge and requiring producers and distributors to register for, and participate in, battery collection for recycling. Each battery recycled means less use of natural resources plus less contamination of the environment.

WHAT YOU CAN DO

- Buy and use rechargeable batteries. While they might be slightly more expensive and don't last quite as long (with each charge) as single-use alkaline batteries, you can recharge them over and over again. In the long run, you'll save money and reduce waste. Rechargeable batteries are also recyclable once they eventually stop working.
- Recycle old batteries and make sure they don't get thrown away in household rubbish. Instead, get a battery recycling box for your home (or office) and then take it to your nearest recycling point – supermarkets often have one. Keeping old batteries in a cool dry place reduces the chances of them leaking and make sure to send them for recycling even if corroded.
- Enlist your kids! Some schools collect batteries and encourage their students to fill boxes for recycling. Recycle your old mobile phone and its battery – some charities collect and recycle old phones to raise funds.
- Avoid buying toys and gadgets that require lots of batteries. Plug-in versions, such as mains-powered electric toothbrushes, are better as they avoid the need for batteries altogether.

RECYCLE FOR GOOD

Some charities collect batteries for recycling and are collaborating with battery recyclers who make a contribution to the charity for every battery collected and returned for recycling. For example, a children's hospice charity in Ireland has diverted millions of batteries from landfill and raised €340,000 as a result since 2011. In Ireland 7,000 retailers have registered to collect batteries for recycling, and WEEE (Waste Electrical and Electronic Equipment) Ireland donates money for every battery recycled.

Additional facts

The manufacture and transportation of batteries emits pollutants into the atmosphere that contribute to climate change. Rechargeable batteries help to reduce these impacts as they can be reused over and over again.

More than 95 per cent of an LED bulb can be recycled, so it is crucial they don't end up in the rubbish.

LIGHT BULB

With satellite images showing the world lit up at night, we live in ever-brighter times. Light bulbs have transformed the way we live and work compared with pre-electricity days. At least 83 per cent of the world's population lives under light-polluted skies. Light pollution disturbs the lifestyles of nocturnal animals and affects the natural rhythms of all life on Earth.

The use of light bulbs enables babies to be delivered safely at night, people to work when it is dark and for us to read in bed without the risk of setting the house on fire.

Thomas Edison is commonly credited with the invention of the first light bulb in 1879; however, a number of inventors had been working on versions of the light bulb since the early 1800s. British physicist Joseph Wilson Swan, for example, received the first patent for a complete incandescent light bulb with a carbon filament in 1879, and his house was the first in the world to be lit by light bulbs.

Both Swan and Edison set up companies to manufacture and sell incandescent bulbs in the early 1880s, and after a few years as competitors they merged their companies and worked together to supply light bulbs and other electrical items.

THE IMPACTS

The most obvious environmental impact of light bulbs is their energy consumption. Twenty per cent of the world's energy use is due to lighting, and in the average household it accounts for 5 per cent of total energy use.

If the electricity powering lighting is largely produced from fossil fuels, it releases carbon dioxide and other greenhouse gases that cause climate change into the atmosphere.

TYPES OF BULBS

It's possible to create light in various ways and each of the types of bulb below use a different technique of illumination.

An incandescent bulb is a source of electric light created by heating a filament. The bulb is made of a glass enclosure with a tungsten filament and it has a base that connects it to the electric current. As an electric current passes through the filament, it heats it to a temperature that produces light. The glass bulb contains a vacuum or inert gas to prevent the filament from evaporating. These bulbs are wasteful, with over 90 per cent of the electricity given off as heat. For this reason, they were phased out of use in Europe between 2009 and 2012.

A halogen bulb is a more efficient and longer-lasting type of incandescent bulb, but these have also been banned in the EU since 2018.

A fluorescent bulb is a glass tube of inert gas with electrodes at each end. Electricity heats up the electrodes, which then give off electrons that pass though the tube to produce ultraviolet radiation. This, in turn, heats the phosphor powder in the tube to make it glow and give off light. Fluorescent bulbs are more efficient (they don't get hot when turned on) and last longer than incandescent bulbs, but can take time to warm up and reach full brightness. Fluorescent bulbs can be long and slim or curved and looped to be more compact.

A light-emitting diode or LED bulb produces light when electrons move around within its structure. So, instead of emitting light from a vacuum (as in an incandescent bulb) or a gas (as in a fluorescent), an LED emits light from a piece of solid matter called a semiconductor. They are long lasting and energy efficient and turn on immediately without any warm-up time. Due to their efficiency and longevity, as well as new regulations in countries ranging from the US and Canada to the EU and China, LEDs are becoming increasingly popular. It is estimated that they will account for 75 per cent of all lighting sales in the US by 2030.

So, less electricity used for lighting means fewer greenhouse gases, which is an important contribution to climate action. Furthermore, because LED light bulbs use less energy, they save money. For instance, it is estimated that if the US switched entirely to LED bulbs it could save $250 billion in energy costs, reduce electricity consumption for lighting by nearly 50 per cent, and avoid 1,800 million metric tonnes of carbon emissions into the atmosphere. Research figures in the European Union estimate that substituting a halogen lamp with one LED could save up to €100 of electricity over the product's lifetime of around twenty years.

Now, walk around your home and add up how many light bulbs you have in total – once you can see just how much money you could save, it makes no sense to not switch to LEDs.

Such energy-efficient lighting could avoid the emission of 12 million tonnes of CO2 in Europe. But, while fluorescent and LED light bulbs may be more energy efficient than incandescent bulbs, they also have more metal-containing components and the natural resources needed to make these need to be extracted from the environment.

These components create an environmental impact when light bulbs are disposed of at the end of their life. Fluorescent and LED bulbs are categorised as hazardous due to the levels of lead, copper and zinc they contain, as well as mercury in the case of fluorescent bulbs. In fact, fluorescent bulbs are three to twenty-six times more damaging than incandescent bulbs (due to the natural resources used in their manufacture and their toxicity), and LEDs are two to three times more damaging than incandescent bulbs if not disposed of and recycled carefully.

Overall, LEDs are best in terms of energy efficiency – just be sure to recycle spent ones.

WHAT YOU CAN DO

- Switch the bulbs in your home to LEDs – this will save you money in the long run and reduce both your carbon footprint and the amount of waste you produce.
- Turn off the lights when you are not in a room to reduce energy use and keep your electricity bills down. Remind any kids to do the same.
- Switch your electricity to a renewable energy supplier, so that the power supply to your home lighting is zero carbon.
- Dispose of old bulbs carefully. If you still have incandescent light bulbs in light fittings, be sure to dispose of them in your normal household waste. Don't put them in the glass recycling as they contain metal.
- Recycle fluorescent bulbs; they should not go in your household waste as they contain chemicals such as mercury. When they are recycled, the metals and glass can be separated and used again. In Europe every fluorescent bulb bought includes an end-of-life recycling fee, so that you can recycle the bulb free of charge at any designated recycling point.
- Recycle LEDs, so that the metals and glass they contain can be reused and so that any toxins are carefully managed. In the EU LED lights are controlled by the WEEE regulations that encourage people to recycle electrical goods. Recolight in the UK helps the lighting industry comply with the WEEE regulations and to date 260 million lamps and their components (including bulbs) have been recycled and used to make new products.

Additional facts

LEDs are the most energy-efficient light bulbs. They don't burn out like other bulbs; instead their brightness diminishes over time until they need to be replaced.

LED bulbs can be used for up to twenty years before they need to be replaced. So, the lifespan of one LED is equivalent to three fluorescent bulbs or thirty incandescent bulbs.

LEDs are six to seven times more energy efficient than conventional incandescent bulbs and cut energy use by more than 80 per cent.

Today 70 per cent of the global population own a smartphone – amounting to a staggering 6.1 billion phones.

There are 3.59 million smartphone users in Ireland and 48.42 million in the UK (2018).

MOBILE PHONE

The rise of the mobile phone has been rapid and far reaching – almost everyone has a mobile phone. From the invention of the first walkie talkies in the 1930s to the launch of the Motorola flip phone in 1989, the mobile phone has revolutionised the way we communicate and access information. But, which is the most sustainable phone?

Across the African continent people skipped the landline stage of telecommunications and went straight to mobile phones, using them to access banking and markets as well as relatives working in cities and overseas. The first smartphone was made by Nokia in 1997 but it was the launch of the touch-screen iPhone in 2007 that made phones user-friendly enough for mass consumption.

Now we use our phones for everything from booking holidays and cinema tickets to doing the grocery shopping and reading the news. They help us to keep in touch with our families and friends whether we call them, share photos and videos through a chat group, or use apps such as FaceTime and Skype to video call. People have been known to sleep holding their phones; it seems we love them more than any other piece of technology.

THE IMPACTS

There is no denying the effects any mobile phone has on the environment. From the extraction of the raw materials needed to make it to the energy involved in its production and use, plus the mountains of waste it generates as we upgrade to the latest model – a mobile phone's impact is significant.

Approximately 7.1 billion smartphones were manufactured worldwide between 2007 and 2017. Between 2007 and 2017 the amount of energy used to manufacture just those smartphones is the same amount of energy required to power India for a whole year. Added to this is the energy required to charge a phone.

According to the United Nations Environment Programme the energy needed to produce a mobile phone produces 60kg CO2e, while charging the phone for a year produces 122kg CO2e – that's enough to drive 2,988 miles (4,809km) in an average car.

By 2020 smartphones will have surpassed laptops and PCs as the main users of energy in this sector. In fact, it is estimated that by 2040 the information and communications technology sector will be responsible for 14 per cent of global emissions, equivalent to the current carbon footprint of the transport sector.

It may look like simply metal and glass on the outside, but inside your phone nothing is so simple. It's packed with a vast array of precious metals. Mining those metals from the ground has a direct impact on the environment as well as the impact of the chemicals used to extract them and the wastewater they produce (called tailings, see box below right).

Iron, aluminium and copper are the most commonly used metals in a mobile phone. Iron is used in the speaker, microphones and in the frame; aluminium is used to make the glass and the frame; and copper is used in the wiring. Gold is also used to make connectors and while gold mining is associated with cyanide and mercury poisoning and rainforest destruction in the Amazon, modern-day prospectors would be better off sourcing their gold from the recycling of phone components rather than the planet.

Another category of components, known as rare earths, are used to make smartphone speakers and microphones, touch screens and the vibrate function. They are extracted using sulphuric and hydrofluoric acids and produce toxic waste as a result.

Cobalt is yet another ingredient needed to make mobile phones, and half of the world's cobalt comes from the Democratic Republic of the Congo where it has been linked to child labour and environmental pollution. According to a 2016 study by Amnesty International, around 40,000 children, some of them as young as seven years of age, work as cobalt miners in dangerous artisanal mines for meagre wages.

The electronic waste (or e-waste) associated with mobile phones is a whole other issue. In the UK a smartphone is used for between twenty-six and twenty-nine months. As new phones are launched, old ones are discarded; built-in obsolescence and software changes push us to change our phones every couple of years.

In developing countries, smartphone recycling often happens in an unregulated market without protections for workers who handle the heavy metals and toxins in the phones. The health impacts on the workers and on their families (via workers' contaminated clothes) has not yet been fully understood.

DISASTROUS CONSEQUENCES

Over forty tailings spills from metal mines occurred between 2007 and 2017 causing pollution and damaging the health of people living in the vicinity. The largest of these was in Brazil in 2015 when an iron ore mine collapsed and released 33 million cubic metres (equivalent to 13,000 Olympic-sized swimming pools) of wastewater into the River Doce, flooding local villages and killing nineteen people.

WHAT YOU CAN DO

- Unplug your phone charger when you are not using it – or switch off at the plug – and save money on your electricity bills.
- Turn down the screen's brightness and use the low-energy mode to prolong battery life.
- Keep your current phone longer. The most sustainable phone is the one you already have. If it is getting slow, update the software, delete files and apps you don't use to help it to perform better. A certified phone refurbisher may also be able to give it a new lease of life.
- Sell or trade in your old phone. Many network providers and phone manufacturers offer incentives to offset the cost of a new one.
- Donate old phones to charity where they are refurbished and sold to raise funds (see also page 91).
- Recycle your phone. The Waste of Electrical and Electronic Equipment (WEEE) Directive was designed to manage the growing stream of electronic waste in the EU, estimated at 12 million metric tonnes in 2020. The directive created collection schemes where consumers return their electronic devices free of charge for recycling and reuse.
- Research before you buy and ask questions about the responsible and ethical nature of a phone's raw materials and the working conditions of the workers who made it. Look at the ethical ratings of mobile phone companies online before you buy.

A SUSTAINABLE SMARTPHONE

Fairphones have been designed with durability in mind and are designed to be repaired, refurbished and reused, contrary to the trend to build in obsolescence. The battery and the camera parts can both be replaced and they work on open-source software that anyone can access. The phones are made with ethically sourced metals and minerals, and the company checks back along the supply chain to ensure that stringent social and environmental standards are met and human rights abuses, such as child labour, are avoided. They also provide good working conditions for staff and collaborate with partners that share their values. The company will help you to fix your phone to get the maximum life from it and is currently developing version 3 of the phone.

Additional facts

It has been estimated that if 10 per cent of mobile phone users unplugged their phone chargers after use, the energy saved over the course of one year could power 60,000 European homes.

By 2040 phone and data centres will have the biggest carbon footprint of the tech industry.

The UN calculated that in 2014 alone the e-waste from small IT devices such as phones was 3 million metric tonnes and less than 16 per cent of this was recycled.

Vinyl sales have sky-rocketed – over 1,427 per cent – since 2007. In 2018, 4 million vinyl LPs were sold in the UK alone, and globally sales reached almost 10 million units.

VINYL RECORD

However you choose to listen to music – vinyl records, cassettes, CDs or streaming – each has its own environmental footprint.

In the early 1900s, the first gramophone records were microgrooved discs that could hold a mere 3–5 minutes of sound. These records were made from a mixture of shellac (a natural resin), wax, cotton and slate. Such records, though, were brittle and prone to water and alcohol damage. Then a shortage of resin during World War II led manufacturers to develop vinyl, or to give it its full name polyvinylchloride or PVC, for short.

Cassettes and CDs followed later and added yet more to the waste footprint of the music industry, as neither format was recyclable, being made of mixed materials. CDs promised to be more durable, last longer and have better-quality sound – but this wasn't always the case and many CDs have been damaged and discarded.

In the digital age we can enjoy high-quality music with no deterioration of sound over time and no physical waste. At the same time, there's been a resurgence in the popularity of vinyl as people seek out the nostalgia, tactile experience and distinctive sound of an analogue record.

THE IMPACTS

A joint study called The Cost of Music by the Universities of Glasgow and Oslo looked into the environmental impact of music over time in the US. The study revealed that in 1977 (the peak of LP sales) globally the music industry used 58 million kg of plastic; in 1988 (when cassette tape sales peaked) it required 56 million kg of plastic; and in 2000 (the peak of CD sales) the industry used 61 million kg. With the transition to digital music, you'd expect the plastic footprint of music to shrink dramatically – and it has, to 8 million kg in 2016.

But the carbon footprint of music is not on the same downward trend – in fact, it's on the rise. Because the streaming of music requires energy to run data centres, retrieve and transmit files and power devices, music-related greenhouse gas emissions have gone up, not down. The research reveals that greenhouse gas emissions associated with music in the US were 140 million kg in 1977, 136 million kg in 1988, 157 million in 2000 and a whopping 350 million kg in 2016. So, the streaming of music has led to significantly higher carbon emissions than at any previous point in the history of music.

Streaming music has a higher carbon footprint but a lower waste footprint than physical music. Physical music requires more resources to produce and contributes to waste, but requires less electricity to play. So, which is the greenest option? Well, it depends on how often you listen to music. If you only listen to a track a few times, it is probably best to stream it, but if you listen to something over and over again it would be worth having a physical copy.

The switch in electricity production from fossil fuels to renewables means that over time the carbon footprint of streamed music will fall. The companies building big data centres are increasingly seeking access to sources of renewable energy when identifying suitable sites. Also the greenhouse gases associated with the electricity to play music at home will also drop off as the grid transitions to renewables.

WHAT YOU CAN DO

- Think about the format you need before you buy – a record, CD or streamed music. If you have it on vinyl, do you need to own it digitally as well?
- Reuse: look for second-hand and pre-loved music.
- Recycle. Vinyl records, CDs and cassettes can't be put in your household recycling bin. The cardboard sleeves on records are cardboard, so they can be recycled, but not the soft plastic sleeves. CDs are generally made of polycarbonate plastic covered with an aluminium coating and are not recyclable. Do recycle the CD cases – just separate out the paper sleeve and plastic case for recycling and remove any soft plastic.
- Switch to a renewable energy provider to cut the carbon footprint of playing music and charging the devices you listen to music on.

THE BEDROOM

An estimated 19 billion pairs of shoes are bought worldwide each year, creating a market worth in the region of US$350 billion in 2017.

SHOES

Almost everyone in the world owns one or more pairs of shoes or sandals, ranging from designer high heels to plastic flip flops.

Sandals made from leather and moccasin-style shoes made from hides and fur are the earliest known shoes; some such shoes date to circa 1600–1200BC from Mesopotamia.

Several innovations over the centuries have enabled the mass production of shoes. First, was the invention of the shoe lace in the UK in 1790; second was the patenting of a sewing machine to sew soles to uppers (the patent was in 1858); and, lastly, the first rubber soles, invented by Irish-American Humphrey O'Sullivan in 1899.

It might be surprising to discover that it wasn't until the end of the 19th century that shoes were made to fit left and right feet!

The invention of glue in the 20th century meant that gluing could start to replace the traditional stitching methods of construction, driving down the cost of shoe production.

The American brand Keds produced the first mass-marketed, canvas-top sneakers in 1917. The word 'sneaker' was coined by advertising agent Henry Nelson McKinney because the rubber sole made the shoes super-quiet.

THE IMPACTS

When it comes to environmental impact, globally the shoe industry is responsible for about 250 million tonnes of CO2e per year. A typical pair of synthetic trainers, for instance, generates 13.6kg of emissions, equivalent to leaving a 13W compact fluorescent light bulb burning for 121 years. According to a 2013 study conducted by MIT, more than two-thirds of the greenhouse gas emissions generated by shoes come from the manufacturing process, which requires many different materials and processes.

Traditionally shoes are made from leather, which up until the late 19th century was tanned using vegetable tannins from tree bark, a process that is still used today to make belts, shoes and soles. However, the dominant method of tanning now is done using chromium compounds – 90 per cent of leather is chromium tanned.

Some people are allergic to chromium, and chromium VI is toxic and carcinogenic. So, in 2015 the EU banned the use of chromium VI in leather footwear, but it is still used in other parts of the world.

Leather is, of course, a by-product of the meat industry, a major source of greenhouse gases, particularly methane, and a cause of deforestation and water pollution. However, making use of the hides associated with meat production avoids waste and maximises the value of each animal.

Synthetic shoes also have a significant environmental footprint. Most of today's trainers are made from synthetic rubber and plastic. The raw material they are made from is oil and, unlike leather, synthetic rubber and plastics (such as polyvinyl chloride and polyurethane) are not biodegradable. Additives to the plastic polymers in synthetic rubber add to the toxicity of the waste and require special treatment.

Natural rubber is rarely used in shoes these days – most of it goes into the manufacture of tyres – but even that has environmental impacts if primary forest is being cleared to grow rubber trees to produce latex. That is why Veja trainers, for example, make sure that the wild rubber from the Amazonian forest they use to make the soles of their shoes is sustainably and responsibly sourced.

SHOES FROM THE SEA

Trainers are starting to be made from recycled plastic, including ocean plastic and ghost fishing nets retrieved from the sea, turning an environmental problem into a new pair of shoes. Adidas have partnered with Parley for the Oceans to make trainers from ocean plastic while a whole range of brands are manufacturing shoes from recycled plastic bottles, including Vans, Rothy's, Indosole, Nike and North Face. Making shoes from recycled materials is a big step forward, and the next will be to close the loop so that every shoe can be recycled and have another life.

The cotton in a trainer or plimsoll's upper and laces also contributes to the consumption of chemicals as conventional cotton production requires large amounts of chemicals (see also page 108). Likewise, the dyes used to colour leather or synthetic shoes add chemicals into the wastestream, all adding to the environmental footprint of the shoes.

The end of a shoe's life is increasingly in focus as plastic shoes are among the most common items found washed up on beaches around the world – flip flops are among the worst offenders. Flip flops are cheap and the only shoe option for the world's poorest people. However, they last only a year or two at most and are then discarded, usually in countries without a waste-management system, meaning they often end up as waste in waterways and, then, the ocean.

WHAT YOU CAN DO

- Decide what's important to you. Choosing the right shoes with the lowest environmental impact is largely informed by personal choice. If you eat meat and are creating demand for beef, then it makes sense to use every part of the cow, including its hide. Advantages are that leather is a natural product, especially if tanned with vegetable tannins, and can biodegrade. Vegan shoes made from organic cotton, old tyres, recycled plastic or natural materials such as jute are now available, as well as vegan leather – though these are essentially plastic.
- Research before you buy – websites such as Rankabrand (see page 251) help you to compare brands and assess their sustainability performance in one place. Or look up the brand yourself and see what sustainability policy and story it has to tell.
- Recycle or donate pre-loved shoes, so that they do not end up in landfill. Some big brands now have their own shoe recycling programmes – such as Nike's Reuse-A-Shoe, which allows you to drop off shoes of any brand for recycling at their retail stores.
- Find a TerraCycle drop-off point near you to recycle old flip flops and rubber shoes. The flip flops are shredded, melted and moulded into new things such as furniture and buckets.
- Organise a shoe swap with friends and cleverly borrow shoes for occasions rather than buying new shoes.

THE FUTURE OF FOOTWEAR?

In 2018 a New York developer and the energy company NRG created a prototype 'footprintless' shoe from recycled CO_2 captured from a power plant. CO_2 effluent from a power plant was captured and liquefied and made into a polymer (plastic) to form a sneaker. The aim was to show that carbon waste from fossil fuel power stations can be repurposed. The shoe is essentially plastic – so steps would need to be added to ensure it can be recycled at the end of its life, to make it truly footprintless.

- Rent shoes. Several online stores allow you to rent designer shoes for special occasions – good for your wallet and for the planet.
- Look out for ethical shoes, such as Sole Rebels made in Ethiopia and sold around the world, which give quality jobs to people living in poverty, recycle fabric and car tyres, use natural, locally grown materials and support the local communities in which they operate.
- Buy locally made shoes and avoid the carbon footprint of transporting them around the world. What could be more luxurious than handcrafted shoes made to fit your feet and last for years?

Additional facts

People in developed countries buy most shoes, with North Americans buying on average seven pairs of shoes a year while people in developing countries are more likely to buy one pair per year.

The majority of the world's shoes are made in China and other parts of Asia.

The estimates of the water needed to grow, dye and process the cotton for just one pair of blue jeans ranges from 2,273 to 8,183 litres.

About 300 million pairs of jeans are made each year.

JEANS

Jeans are the most popular item of clothing in the industrialised world, with the average American owning seven pairs of jeans. About 70 million pairs of jeans are sold in the UK every year, while in warmer climes down under, Australians buy fewer than 10 million pairs.

The name 'denim' comes from serge de Nîmes – a strong material made from wool. By the 1700s, it had become a mix of wool and cotton and was used to make sails. Some Genovese sailors thought the sail fabric would make great trousers and soon this new type of workwear was born. The name 'jeans' emerged 100 years or so later when the cloth, typically dyed blue or brown to hide the dirt, was solely made of cotton. At that stage both the textile and the casual cotton or denim trousers made from it were called jeans.

Classic jeans as we know them today – hardwearing garments made from indigo-dyed denim (for the blue) with pockets and rivets – were patented by two Americans (Jacob Davis and Levi Strauss) in San Francisco in 1873. While the additions of belt loops, zips and decorative topstitching were added over time, the basic fabric – blue on the outside, white on the inside – remains. Today, jeans often have spandex added to the weave to allow for stretch and synthetic indigo is used for dyeing.

THE IMPACTS

A life cycle analysis of jeans to determine their environmental impact reveals some shocking facts. According to a study in 2013 by Levis, the amount of water used to make a pair of 501 jeans – including the water needed to grow the cotton and the water needed to make, dye and finish the fabric – is estimated to be 3,781 litres. Most of the water, 68 per cent, is used growing the cotton and the next most important factor is washing the jeans over their lifetime – responsible for 23 per cent. The growing of cotton also contributes to water pollution and eutrophication associated with excessive levels of nitrogen and phosphorus from fertilisers. The other impacts from the Levi's study revealed that one pair of jeans produces 33.4kg CO2e, (equivalent to driving 69 miles (111km) in an average US car), has a eutrophication effect (the total amount of phosphorus) of 48.9g PO4e (phosphate equivalent) as well as occupying $12m^2$ of land a year to grow the cotton.

The indigo used to dye jeans also has an environmental impact. Most jeans are made in Asia where it is estimated that 70 per cent of rivers and lakes are contaminated by over 9 billion litres of wastewater produced by that continent's textile industry.

An investigation carried out by Greenpeace in 2010 in China tested the outflows near dyeing and finishing facilities in denim-producing towns and found five heavy metals (cadmium, chromium, mercury, lead and copper) in seventeen out of twenty-one samples of water and sediment.

In the last few years, several innovations have reduced the water used in the indigo dying process by 20–50 litres per garment and reduced the amount of chemicals and salt required. Nowadays the sodium hydrosulphite traditionally used in indigo dyeing can be replaced by an organic agent to dramatically reduce the amount of chemicals used by 70 per cent and to eliminate salts completely. Fewer chemicals in the process means less water treatment is needed along with more potential for reusing the water.

The finishing of denim to give it a distressed look also uses chemicals and water. New techniques use lasers and ozone technologies instead of the pumice stone, sandpaper and potassium permanganate used in the past to age the denim. According to Alex Penades of Jeanologia, who has been behind the innovations to reduce the environmental footprint of jeans, in 2015 only 16 per cent of the jeans in the world were made in a sustainable way (see pages 112–113); this had risen to 35 per cent by 2018.

Jeans also have a carbon footprint – largely arising from the energy used to make them. According to Levi's life cycle analysis, each pair of jeans has a carbon footprint of 33.4kg of CO2e, – equal to watching a large-screen plasma TV for 246 hours. Of this, 37 per cent is due to washing and drying the clothes, 27 per cent to making the fabric, 9 per cent to growing the cotton, 11 per cent to transport, 9 per cent to sewing the jeans and the remainder to packaging and waste.

So, an important way to reduce the carbon footprint of your jeans is to wash and dry them less often. Americans use most energy as they typically wash their jeans more, wash them on cold cycles and then dry them in a tumble dryer. Europeans, on the other hand, are more likely to line-dry their jeans, reducing the carbon footprint of the jeans. Nevertheless, wearing jeans ten times before washing them could reduce energy use by 75 per cent.

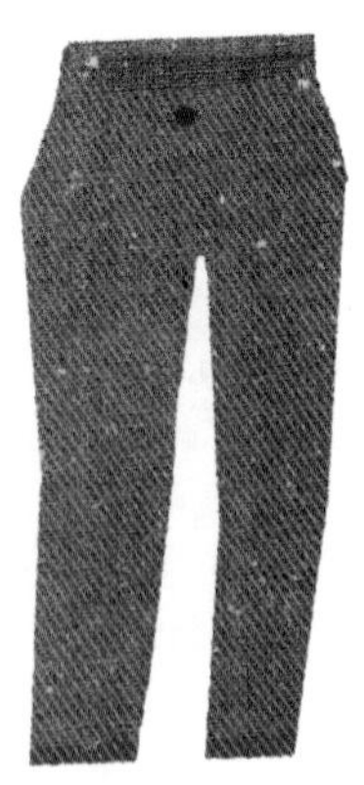

Most jeans are made of a variety of materials in addition to the main constituent of denim. Plastic and metal buttons, zippers, metal rivets, leather or plastic branding on the belt and synthetic labels with washing instructions feature on almost every pair of jeans. Reducing the number of materials jeans are made from would make them more recyclable.

WHAT YOU CAN DO

- Wash your jeans less often (ten wears to one wash), on a cool wash, preferably on an eco-setting if your washing machine has one, and line-dry them to dramatically reduce the carbon footprint of your jeans, as well as the amount of water used. For example, if you wash your jeans once a week in an energy-efficient, front-loaded machine on a cold cycle and line-dry them, the carbon footprint is about 2.58kg CO_2e, while it is a whopping 9.92kg CO_2e, if washed in warm water in an energy-efficient machine and tumble dried. If you have an older inefficient machine and use a dryer, the carbon footprint rises to 14.5kg CO_2e, – almost six times the carbon footprint per wash!
- Make do and mend. Repair holes in jeans and sew buttons back on to extend their life. If you aren't good with a needle and thread yourself, find a local repair shop (many dry cleaners offer a repair service).
- Look for pre-loved jeans and jeans made from used jeans.
- Give old jeans another life – make old jeans into shorts or skirts. Take them to a charity shop or jeans recycling drop-off point for recycling. Some brands of jeans have recycling points within their shops.
- Do your research before you buy – look for sustainable brands that use organic and fairtrade cotton and check their ethical and sustainability policies. All this information should be transparently available on a brand's website. Check their environmental impact, see if the jeans can be easily recycled and make sure that the company supports good working conditions and fair wages for all of its staff.
- Avoid jeans with spandex, polyester pockets, glitter, decorative buttons etc. as it makes them even harder to recycle and the plastic parts will never biodegrade.

Additional facts

The Saitex factory in Vietnam saves 430 million litres of water – equivalent to the annual water consumption of 432,000 people – by using recycled water and closed jet-wash systems when giving jeans that worn look. As a result, Saitex denim uses 1.5 litres of water compared with 80 litres of water for the same amount of denim made using standard practices.

Drying finished jeans in factories by aerial drying (hanging them up to dry in the factory) instead of using tumble dryers can reduce the CO_2 emissions of jeans manufacturing by 80 per cent.

A discarded fleece will take hundreds of years to break down and even then will only break down into smaller pieces, or microfibres, that can hang around in the environment for thousands of years.

FLEECE JACKET

Once worn only by the adventurous who climbed mountains and ventured to the poles, the fleece jacket is now ubiquitous in any country that feels a chill. Lightweight, warm, easy to wash and quick to dry, fleece is a revolutionary fabric that performs like wool but with the added benefits of not holding water, staying warm even when wet and not being itchy. Fleece jackets are now sold on every high street around the globe, often very cheaply.

Fleece is a form of polyester – a durable, synthetic fabric made from non-renewable petroleum derivatives, or in other words oil (see also pages 134–135). In 2016, 65 million tonnes of plastic was produced to make textile fibres.

Fleece has proven to be so popular as a lightweight, warm material that it is now used for all sorts of garments from jackets, hoodies and trousers to blankets, hats and gloves.

THE IMPACTS

That warm and lightweight jacket may not cost that much in money but it has a serious environmental cost. When you wash fleece, tiny fibres escape down the drain or through the filter in your washing machine and make their way into the water system. In 2016, researchers at the University of California at Santa Barbara found that, on average, synthetic fleece jackets release 1.7g of microfibres each time they are washed. They also found that older jackets shed almost twice as many fibres as new ones. When these microfibres travel from your washing machine to your local wastewater treatment plant, up to 40 per cent escape into rivers, lakes and the sea (depending on local wastewater treatment conditions). As a result, these microplastics are found in the water we drink, the soil we cultivate and the food we eat. They are so light they easily become air borne and, as a result, are even in the air that we breathe.

In fact, plastic microfibres represent 35 per cent of all microplastics found in marine environments and have been found in all ecosystems across the globe. In aquatic environments, fish and other marine and freshwater animals mistakenly eat them or feed them to their young.

Research from Australia, with similar findings now seen around the world, has found plastic in animals all along the food chain from zooplankton to seabirds, with worrying signs that plastics act as a vehicle to transport toxins and metals, such as lead, cadmium and arsenic, into animal tissue.

Scientists have also found that microplastic particles in fish can physically damage their organs and may leach hazardous chemicals that can affect their immune function, growth and reproduction.

WHAT YOU CAN DO

- Choose natural fibres, such as wool with its naturally water-resistant oils, over synthetic textiles. Wearing fine wool layers made from merino wool will dry almost as fast as fleece but have the added benefit that they get less smelly than synthetic textiles.
- Launder less. Wash a fleece jacket as little as possible to reduce the amount of microfibres released. Choose a low-temperature (30°C) synthetics wash programme (hot water promotes the shedding of microfibres) on your machine with a short and slow spin cycle that will reduce the numbers of microfibres released.
- Line-dry synthetic clothes, since the action of a tumble dryer and its heat cause microfibres from synthetic clothes to be released.
- Catch the microfibres. You can buy a mesh bag (such as a Guppy Friend) for your washing machine that claims to catch 86 per cent of the fibres shed from synthetic textiles. You can then dispose of the caught fibres in your bin.
- Choose recycled fleece instead. Some manufacturers now produce fleece from recycled plastic bottles and recycled ghost fishing nets (those lost at sea; see also page 210). This gives plastic waste a new lease of life and these fabrics perform just as well as other fleeces.
- Get as much use as possible from any fleece that you already own.
- Hand them back. Some outdoor brands accept old or unwanted fleece clothes and find new uses for them in disaster relief or microenterprises.

Additional facts

A single fleece jacket sheds as many as 250,000 synthetic fibres during laundering.

Guppy Friend bags were invented by two German surfers and are designed to catch microfibres when you wash clothes.

While it is technically possible to recycle fleece into something else (such as carpet, for example), such recycling facilities are not widely available and the number of times synthetic fibres can be recycled is limited. And if textiles are made from a mix of synthetics (such as polyester and nylon), they can't be recycled at all.

Every year 2 billion pairs of tights are produced, worn once or twice and then thrown away.

TIGHTS

These days you can pick up a pair of tights with your weekly shop and even in some newsagents, but they weren't always so easy to find. My granny told me stories about the difficulty of getting tights during the war years when she was a young woman, and the lengths women went to 'paint on' stockings, even adding a seam line with an eyeliner.

Nylon, a recent invention in the 1930s, had only just started to be used to make tights in 1940. But they were an instant success – with 64 million pairs of 'nylons' sold in the US in their first year of production alone. So, when nylon was prioritised for use in the war effort to make parachutes and the like, women's hosiery had to take a back seat.

THE IMPACTS

Nylon thread is a strong and light plastic and when you add elastane you can create a stretchy, thin fabric to cover the legs. Making nylon involves the extraction of natural resources, in this case petroleum, and the use of energy to manufacture the fabric, which results in emissions and the release of air pollutants, such as nitrogen dioxide. This greenhouse gas contributes to climate change (and is 300 times more powerful at causing global warming than carbon dioxide), depletes the ozone layer and creates smog.

Although nylon absorbs less water than a natural fibre, such as cotton, large volumes of water are required, nonetheless, for the cooling stage during its production and for dyeing, with the added risk of water pollution associated with the chemicals and dyes.

Nylon tights are not biodegradable, yet many of them, especially the thinner ones

(10–15 denier), are so fine that they are usually worn only once or twice before they get a ladder and have to be thrown away. Tights are not currently recycled, so they end up in the bin and then in landfill or incineration. They are essentially a disposable fashion item and one of the worst offenders of fast fashion (see also pages 112–113).

But nylon is actually recyclable – we just don't have the recycling services commonly available yet. New efforts to recycle nylon to make tights (see right) and swimsuits (see page 210) means that more sustainable nylon fabrics are on the way.

WHAT YOU CAN DO

- Be careful with the tights you have. Handwash delicate ones to extend their life (or use a mesh bag to protect them within the machine) and darn the toes of thicker opaque ones to help them last longer.
- Consider opaque tights made from cotton or wool instead of nylon. Both wool and cotton tend to be more durable and if they are good quality they should wash well without getting bobbly. Bamboo is also emerging as another natural fabric alternative to nylon, with some reports, though, that bamboo tights can shrink when washed.
- Buy sustainable tights – read all about it in the box, right.
- Invest in a pair of silk tights – the only real alternative to sheer nylon tights. They are pricey, so you'll need to treat them well and get lots of wear out of them. You might want to look for 'peace silk', which does not kill the silkworm in the process of extracting the silk.
- Get brave and go bare legged or wear cotton leggings instead of nylon tights – if the look allows!

SUSTAINABLE HOSIERY

Linn and Nadja were frustrated by the dominance of throwaway, cheap and polluting tights on the market. Convinced there had to be a better way to make tights, inspired by the luxury tights of the past and the technology of the future, they set about making sustainable tights. And their company Swedish Stockings was born.

One hundred per cent of the nylon and elastane used to make this company's tights is sustainable. The nylon comes from consumer waste and waste from factories, for example, those that make sports clothes. At the moment, the elastane comes from excess waste from virgin elastane production. This is because the technology to separate elastane from other materials in old clothes so that it can be recycled is still being developed.

The company uses environmentally friendly dyes in their factories, treats the water used to ensure it is clean and safe and uses solar power to run the manufacturing process. They also take back old tights for future recycling when the technology is available to separate the nylon from the elastane – they believe this time is not far off.

They also offer advice on how to wear and care for your tights to make them perform best for you and the environment, including to only wash tights every five to six wears (to reduce your environmental footprint and to protect the fabric), to handwash tights in cold water with very little detergent or on a delicate cycle in a Guppy or mesh bag (see also page 109) in the washing machine and to avoid using fabric softener as it damages and weakens the elastane.

FAST FASHION

We all shop. All the time. It is almost unavoidable in the modern world. The shopping that a lot of people enjoy most, and that is a leisure pursuit for some, is clothes shopping. We consume 400 per cent more items of clothing than we did 20 years ago and the predictions are that this will only continue to increase up to 2030.

People in the UK buy more clothes per person than in any other country in Europe, yet many of those clothes are worn only once or twice and sometimes not at all. As a result, UK citizens discard about 1 million tonnes of textiles a year, equivalent to £140 million worth of clothes per year. In fact, the first thing that strikes you when you research the fashion industry and its impact on people and the planet is the sheer scale of the numbers involved (see opposite). Whether it is the size of the industry, the scale of the pollution or the injustice of the conditions workers endure, the numbers are just staggering.

KEEPING UP WITH THE JONESES?

So, why do we buy so many clothes? It used to be that the fashion industry produced twice a year: spring/summer and autumn/winter. But now there are multiple collections per year, with pre-season, cruise, resort and party collections all added, catering to the wealthy but also influencing the high street. Most clothes sold by high street retailers have a shelf life of twelve weeks maximum and then they are discounted for clearance. In 2019, a report by the Environment Audit Committee of the UK House of Commons into fashion, consumption and sustainability heard evidence that the luxury brand Burberry burned £28.6 million worth of clothes and cosmetics in 2017/18 rather than have unwanted stock sold cheaply. Clearly things are out of control.

THE PRICE OF CHEAP CLOTHES

Fast fashion is the term given to the modern accelerated fashion business model that has multiple fashion collections each year, a quick turnaround and lower prices. It is constantly offering new products to meet consumer demand and it has created a new norm – where clothes are cheap, not designed to last and, as a result, are worn just once or twice and then discarded. These days, you can buy a dress for £5 and a T-shirt for £2 – they are essentially disposable.

Research suggests that 17 per cent of young people will not wear an outfit again if it has been snapped and seen on Instagram. Such an approach has started a vicious cycle: fashions change constantly so people buy more; and as they wear clothes for less time, they can be made to a lower quality and hence cheaper, so they can buy more. It all drives the costs of clothes ever downwards, makes them disposable and reduces margins so that corners get cut on workers' conditions and environmental controls. For example, the price of paying just £5 for a new dress made in the UK is that the workers who made that dress earn as little as £3 an hour, well below the national minimum wage.

FASHION BY NUMBERS

The fashion industry is a **US$3 trillion** global market • This sector employs an estimated **60–75 million** people worldwide • Eighty per cent of them are women between the ages of 18 and 35 • Garment workers earn only **1–3 per cent** of the price paid for an item of clothing • The fashion industry produces **3.3 billion tonnes** of CO_2e per year – close to the carbon footprint of the entire European Union (28 member states) • It uses one-quarter of all the chemicals produced worldwide and is responsible for **20 per cent** of industrial water pollution globally • Producing textiles is second only to agriculture in the amount of water it consumes • One pair of jeans and one T-shirt consumes **20,000 litres** of water (growing the cotton, manufacturing and consumer use) • **100 billion** items of clothing are produced worldwide each year, and three out of five are thrown away within a year.

THE IMPACTS OF OUR INSATIABLE CONSUMERISM

ENVIRONMENTAL IMPACTS – Everything from the water consumption associated with cottons, the pollution associated with dyeing and treating textiles and the petrochemicals and microplastic pollution associated with making and washing synthetic clothes.

SOCIAL IMPACTS – Around the world garment workers are underpaid. From Leicester in the UK, where workers are paid below the minimum wage, to Bangladesh and other Asian countries, where workers are not allowed to negotiate their pay and conditions. In fact, a 2016 report found that 77 per cent of 71 leading retailers in the UK reported that there was a likelihood of modern slavery occurring along their supply chain and bonded labour is known to be used in the cotton industry in Uzbekistan and Turkmenistan, both of which are in the top 10 cotton-producing nations.

WASTE IMPACTS – Buying more clothes means more clothing waste. Approximately 300,000 tonnes of clothing ends up in bins in the UK every year, with 20 per cent of this going to landfill or incineration. Waste is also generated in the production of clothes where at the cutting stage as much as 15 per cent of the fabric goes to waste. It's not waste until it's wasted.

We need a new approach to fashion where we share, repair, swap and rent clothes and shoes instead of buying more and more new stuff, for instance:

- Rent or swap clothes for special occasions from job interviews to weddings from Depop, YCloset, Yeechoo, Rent the Runway and NuWardrobe.
- Organise a clothes swap with friends. So much fun and everyone leaves smiling with something new to them to wear.
- Learn how to repair and look after your clothes so that they last a lifetime.

Each year, the US government estimates that over 136 million kg of single-use, plastic dry cleaning bags wind up in America's landfills.

DRY-CLEANING COVER

The increase in the public's awareness about single-use plastics means that many people and dry-cleaning businesses are now looking at the volume of plastic wrap used to cover dry cleaning and questioning its impact.

Dry cleaning is actually in decline as people wear clothes made from more low-maintenance fabrics, such as cotton and synthetics.

As well as the plastic bags that envelop the clothes, dry cleaning uses solvents to 'dissolve the dirt' from garments. Clothes dropped off for dry cleaning are sorted according to colour, fabric or the type of stain. They are then pre-treated using a spray before being cleaned in a giant washing machine that uses solvents instead of water. At the end of the cycle, high temperatures evaporate the solvents and the clean clothes are ironed, put on a hanger and covered in a plastic bag ready for collection.

THE IMPACTS

So, why do we use dry-cleaning covers in the first place? They maintain dry-cleaned clothes in top condition, keep off rain and prevent spills or stains on the journey home, plus they keep off dust while in the wardrobe. But overall these bags are not a good storage solution, they can trap moisture and dry-cleaning chemicals in with your clothes and can cause mildew growth, which damages the fabric. Experts in clothes storage say that leaving dry-cleaned clothes in the plastic cover can even cause yellowing and weakening of fibres – they're best stored out of the plastic.

Dry cleaning is a heavily regulated industry due to the fact that the solvents used are damaging to the environment and need to be treated with great care. The original dry-cleaning technique, developed in the 1850s, used kerosene but this was replaced by perchloroethylene or PERC in the 1930s. Today's modern machines use 30 per cent less PERC and are much more carefully regulated.

PERC has been found to be both neurotoxic and carcinogenic. It has been linked to an increased incidence of cancers in dry-cleaning staff, and it affects the air quality in and around dry cleaners.

As a result, environmental regulators place controls on the use of PERC and other dry-cleaning solvents and are inspected regularly.

WHAT YOU CAN DO

- Dry clean less – think before you drop things off after one wear. Does it really need to be cleaned? If odours are the issue, natural fibres benefit from airing outdoors. Ironically, quality clothes made from natural materials may require dry cleaning where disposable fashion made from synthetics doesn't. So, weigh up the pros and cons.
- Look out for environmentally friendly dry cleaners – more and more dry cleaners are offering PERC-free cleaning techniques, using liquid silicone from sand or a CO_2 blasting process. Green Earth dry cleaners operate around the world with a commitment to sustainability and PERC-free cleaning. The dry-cleaning solution these dry cleaners use contains silicone as a non-toxic, non-hazardous alternative.
- Take your own reusable suit bag or cover to the dry cleaners – put your name on it so that you make life easier for the dry cleaners.

CLOTHES HANGERS

Many hangers used to display clothes in shops are single use, discarded when the item is bought. Wire hangers are cheap and can be recycled at the end of life as they are made of steel. Wooden hangers tend to last a long time but the source of wood is important; best to buy FSC-certified hangers. Plastic hangers are not generally accepted in household recycling as they can be made from a mix of plastic and metal. With any hanger, but especially plastic ones, reuse is key – whether that be by you, a charity shop or the retailer itself.

Additional facts

An estimated 15 billion hangers are used globally every year and some last only three months.

In 2018, M&S stores returned over 100 million hangers to Braiform (a company that makes hangers and organises their reuse) for recycling or reuse, which if converted into carbon, would be the equivalent to taking nearly 5,000 cars off the road for that year.

Arch & Hook is providing a game-changing innovation in the global hanger industry – making hangers that are sustainable from an ecological and an economic point of view. It makes long-lasting hangers for retail and hospitality businesses using FSC-certified wood. Its latest hanger is made from recycled ocean plastic, can be used for any type of clothing and is fully recyclable.

Only 5 per cent of household textiles – tea towels, sheets, duvets and curtains – are collected for reuse or recycling in the UK.

A duvet made from recycled polyester contains the equivalent of 120 plastic bottles.

DUVET

Throughout life you'll spend, on average, twenty-six years sleeping and seven years trying to get to sleep – a total of thirty-three years under a duvet.

THE IMPACTS

Synthetic duvets are made from polyester, a fabric made from petroleum that is energy-intensive to make and not biodegradable; it is also less breathable than natural fibres. A good synthetic duvet should last at least five years while a down duvet can last twenty, thirty, even forty years if it is good quality and looked after.

Down is a natural product that comes primarily from ducks and geese and is biodegradable. Controversy in the late 2000s about live plucking of geese for their down and the cruelty involved led industry coalitions in Europe, China and Canada to ban the practice in favour of plucking after the animals have been slaughtered for food.

Duvets can also be filled with other natural fibres such as wool and cotton, with more manufacturers using organic and responsibly sourced materials. Synthetic duvets using recycled polyester filling are also now available.

WHAT YOU CAN DO

- Look after the duvet you have. Simply airing a duvet outside in the sunshine is enough to sanitise the cotton cover and evaporate any moisture in the filling, taking bacteria, toxins and dead skin with it. Spot clean any spills or stains to reduce the frequency of washing and drying the whole duvet.
- Look for brands that are transparent about the sources of their materials, their sustainability policy and their commitments to animal welfare.
- Pass on old duvets to charities if in good condition or to animal shelters if not.

It's not known how many earplugs are used per year, but we can look at the numbers involved with an example. Take a company with 200 workers, who use two pairs of foam earplugs per day. Each year would see 100,000 pairs of earplugs sent to landfill. If that's from just one company, you can only imagine the horror of scaling it up.

EARPLUG

Whether you use earplugs to sleep, swim or to protect your ears from noise at work – you will appreciate their effectiveness.

Polyurethane (PU) earplugs (also known as foam earplugs) are soft, comfortable, easy to insert and available in a variety of different shapes, sizes and colours. PU foam is also available in varying densities, so different earplugs can be used for different levels of noise.

THE IMPACTS

Most earplugs are made from PU, which is soft and easy to insert and can be used for heavy-duty noise or just the odd snore. However, PU is not recyclable and earplugs made from PU end up in landfill and incineration or, worse still, as litter.

Earplugs can also be made from silicone but these typically have a lower level of noise protection, though they work well for swimming. Silicone is essentially a plastic so these earplugs pose the same risks in the environment as PVC or PU earplugs.

Less common but less impactful are wax earplugs made from cotton-encased wax; they are made from natural ingredients, such as beeswax, lanolin and cotton, so they are 100 per cent biodegradable. They work well for swimmers and for moderate levels of noise and, like silicone, they are reusable.

WHAT YOU CAN DO

- Choose wax earplugs over silicone or PU as these are biodegradable.
- Consider custom-made earplugs made from medical-grade silicone that are reusable and fit your ear perfectly. They are more expensive but one pair should last at least one year.

THE BATHROOM

A person uses on average 300 toothbrushes during their lifetime (an average of four toothbrushes per year), equating to almost 5.5kg of plastic waste.

Worldwide, more than 1 billion plastic toothpaste tubes are thrown out each year.

TOOTHBRUSH, TOOTHPASTE AND DENTAL GADGETS

People have always sought ways to get food out of their teeth and combat bad breath. As far back as 3500BC, Babylonians were using carefully selected twigs to brush their teeth – and this tradition lives on today in many parts of the Middle East, Africa and Asia.

The twigs of trees such as orange, lime, neem, tea tree and *Salvadora persica* are used as chewing sticks in many parts of the world; they have antibacterial properties, contain fluoride and help to keep breath smelling sweet. The World Health Organization recommends the use of chewing sticks in countries where plastic toothbrushes are beyond the means of many people, and research has found that they are, in fact, as effective in ensuring oral hygiene as a toothbrush and toothpaste.

Chewing sticks are a far more appealing option than the early toothbrushes, which were made from animal bone handles with pig hairs for bristles. The ingredients used in early toothpastes sound as if they would do more harm than good in terms of fresh-smelling breath. Made from ox hooves, egg shells, pumice, oyster shells, charcoal and bark – you'd be hard-pressed to choose an option to try. The first commercial toothpastes were sold as powders, with pastes packaged in glass jars becoming the norm in the late 1800s. The first fluoride toothpastes to prevent tooth decay were developed in 1914.

THE IMPACTS

Toothbrushes are now overwhelmingly made of plastic, unless you live in a society where the chewing stick is still the preferred or more affordable option. The bristles are also plastic in the form of nylon and the grip is generally a type of synthetic rubber, also plastic. So, although a typical toothbrush is only used for a few months, it is made to last many lifetimes, which is a problem if they are not disposed of properly – in fact, toothbrushes are among commonly found pieces of marine litter on tropical beaches.

Toothbrushes are not recyclable as they are made of different types of plastic, and electric toothbrushes are treated as electronic waste, so those need to be recycled just as you would an old toaster or kettle. Toothpaste is typically packaged in a plastic tube and, since it is impossible to get the tube completely empty and clean, it is not recyclable. Plastic microbeads made from polyethylene have found their way into some brands of toothpaste promising whiter teeth. Up to 1.8 per cent of the weight of a tube of toothpaste can be made up of microbeads (see also page 137). Microbeads have now been banned in countries including the UK, US and New Zealand.

Most modern toothpastes contain fluoride to prevent tooth decay. Over time fluoride has also been added to water supplies as a public health measure. This double-pronged approach has led to concerns that people are now getting too much fluoride. If children ingest too much fluoride as their teeth are forming they can become discoloured and stained. Public health officials are weighing up the risks and benefits of adding fluoride to drinking water, comparing the benefits to children's teeth from having this fluoride with any potential for overexposure to the mineral. Most drinking water in countries such as Ireland, Australia and New Zealand is fluoridated. Only 10 per cent of drinking water in the UK has added fluoride.

Some toothpastes contain the antibacterial agent triclosan, and there are fears that its overuse could lead to bacteria (in humans and in animals) becoming resistant to its effects. In terms of impacts on people, there is no conclusive evidence of harm. Nevertheless, Colgate Total toothpaste has been relaunched with a new formula that is free of triclosan; Procter & Gamble and Unilever no longer use it in any of their products and Johnson & Johnson no longer uses it in its baby, cosmetic and personal care products.

DO YOU FLOSS? TIME TO SWITCH

If you use dental floss or dental floss sticks, the floss part is made from nylon, PTFE (see page 39) and synthetic wax and the sticks are plastic. Neither floss nor floss sticks are recyclable and as they are all essentially made from oil, are plastic and are not biodegradable. They also tend to come in a plastic container with a metal cutter, and so are not recyclable. Never flush floss down the toilet; it won't biodegrade and may not be captured by the wastewater treatment plant. Out in the world, stray floss poses a threat to aquatic life that can become entangled or ingest it. But flossing is an important part of dental hygiene – so try a wooden toothpick or some silk or bamboo floss instead.

WHAT YOU CAN DO

- Say goodbye to your plastic toothbrush. Try a bamboo toothbrush. Just remember to cut off the nylon bristles before you compost the rest of the brush when you need to change it. If home composting, cut the brush handle into small pieces – you will need a small saw for this – before putting it in the compost heap to speed up decomposition – or repurpose old toothbrushes as stakes or as plant labels in your garden.

- Use rechargeable batteries in your electric toothbrush if it's battery operated to reduce waste; of course, be sure to recycle old batteries separately.
- Look for toothbrushes made from plant-based plastics that can be industrially composted or a recyclable plastic toothbrush made from recycled plastic.
- Don't leave the tap running when you are brushing your teeth as this just wastes water. If you leave the tap running while you brush your teeth for 2 minutes – that's about 12 litres of water down the drain!
- Switch to a plastic-free floss alternative – such as silk or bamboo floss with beeswax in a reusable container.
- Avoid floss sticks – these are usually made of plastic and are single use. Swap for a bamboo version or floss without a stick.
- Examine the ingredients in toothpaste before you buy it to avoid microbeads and triclosan. If you are concerned about fluoride opt for fluoride-free toothpastes made from natural ingredients such as peppermint, tea tree oil and aloe vera. If making a change, check with your dentist first because if there's no fluoride in your water you need to have fluoride in your toothpaste.
- Recycle the plastic lids of toothpaste tubes.

PLASTIC-FREE TOOTHBRUSHING

Canadian company Bite makes toothpaste and mouthwash tablets from natural ingredients to reduce the waste associated with toothpaste tubes and makes the tablets from all-natural ingredients so that they are good for people and the environment. You chew the tablet to make a foam in your mouth and then brush as usual. The tablets come in glass jars so the whole product is plastic free. Discuss with a dentist before making any changes.

- Avoid plastic tubes of toothpaste if you can and look for toothpaste in aluminium tubes that come with a winder so that you can get all the toothpaste out and then recycle the empty tube (if the brand you buy doesn't have a winder to ensure the tube is empty, don't put it in the recycling as it will cause contamination). Another option is toothpaste or toothpaste powder, made of bicarbonate of soda and essential oils and packaged in a glass jar, and toothpaste tablets. Check with your dentist as some are only suited to occasional use.
- Try a chewing stick for a zero waste option.

Additional facts

You spend 38.5 days of your life brushing your teeth.

The mean annual usage of toothpaste for a family of four in EU member states in 2008 was ten tubes a year – none of which is currently recyclable.

THE FIRST BATTERY-FREE 'ELECTRIC' TOOTHBRUSH

A few years from now you might have swapped your disposable plastic toothbrush for a long-lasting, battery-free 'electric' toothbrush.

The toothbrush runs on renewable energy that you generate when you twist the base of the handle to wind the spring motor inside. The motor produces as much power as an electric motor and two twists are enough to power 80,000 brushstrokes.

The brush is designed to meet American Dental Association standards, so that it puts just the right amount of pressure on your teeth and gums to clean them without causing harm.

Called the Be., the toothbrush is designed to last for ten years or more and is fully recyclable at the end of its life.

The fact that the toothbrush doesn't need batteries or cables to recharge it means zero emissions charging, fewer raw materials and no battery waste. There is also the convenience of not needing to pack chargers when you are travelling.

The inventors say the toothbrush is made from 90 per cent recycled plastic and is 100 per cent recyclable. Only the minimalist toothbrush heads need to be replaced every couple of months and they are made from a plant-based plastic made from starch and bamboo.

The brushes are delivered in recyclable cardboard tubes and replacement brush heads are shipped in paper envelopes to minimise packaging waste.

Soap played a pivotal role in one of the greatest breakthroughs of modern medicine – Charles Lister's discovery that using soap to wash hands reduced the spread of disease. Soap is a real hero... but we seem to have lost sight of its benefits and instead have enshrined it behind layers of plastic.

LIQUID SOAP

Historically, the Ancient Babylonians made soap from a combination of boiling oil with ashes while the Egyptians preferred salt with a mix of plant and animal oils. Commercially available soaps have come a long way since then, but the advent of liquid soaps and their impacts on the environment have many people returning to the humble bar of soap.

Liquid soap is more expensive than a bar of soap and lasts considerably less time. Few people are content with a single squirt, so a huge amount of soap is simply washed down the sink. Washing your hands with liquid soap is estimated to use over six times more soap than using a bar. So, if the average person uses the equivalent 656 bars of soap in their lifetime, we can estimate that this probably equates to 3,000–4,000 bottles of liquid soap. That's a lot of soap down the drain and a mountain of plastic bottles.

Soap is made of plant or animal fat and some fragrance, but beyond that, the list of ingredients can be endless. A good rule of thumb is that if the list is long and full of names and numbers you don't understand, steer clear and choose a bar of natural soap with a short and comprehensible ingredient list.

The chemicals present in many mass-produced liquid soaps are of questionable benefit to our health or that of the creatures that live in the waterways and oceans where these chemicals ultimately end up.

THE IMPACTS

There are many downsides to liquid soap. Liquid soap is heavier than a bar of soap because it contains more water, so it's more expensive to transport in fact. Friends of the Earth estimates that the carbon footprint of liquid soap is 25 per cent more than a bar of soap. Liquid soap also takes fives times more energy to produce and can use up to twenty times more packaging.

What's more, the bathroom is often a black hole when it comes to recycling plastic bottles – one study showed that only 20 per cent of Americans recycle bathroom items.

WHAT YOU CAN DO

- Buy soap 'naked' or wrapped only in paper – no plastic required.
- Opt for the natural choice. All soap contains oils – vegetable or animal. Vegetable oils tend to have a smaller environmental footprint but avoid palm oil (see also page 22) because its production is associated with deforestation. A natural soap based on olive oil with lavender or natural essential oils are kind to your skin and to the environment.
- Use every last bit. Recycle the ends of bars of soap into a new bar of soap. By putting bits of leftover soap in water overnight and then heating the mixture over a low heat, stirring carefully until the soap dissolved, you can revitalise the bits of soap that could end up in the bin. Add a little vegetable oil (1 dessertspoon per cup (about 240ml) of mixture) and some essential oils and food colouring if you wish. Pour the soap into a silicone mould (such as a muffin case) and leave to solidify. Then, you have new soap! There are plenty of ideas and instructions on how to do this online.
- Choose bar forms for all personal products. It is possible to buy face wash, shampoo and conditioner in bar form – hello plastic-free bathroom!
- Refill rather than buy new. If you can't bear the idea of giving up liquid soap, cut down on the need for recycling or waste by having a refillable pump container, so that you can reuse that over and over again. Source large containers (to reduce packaging) and make sure the empty, clean and dry pump bottle makes it into your recycling bin when you've eventually finished using it. Rinse out the little bit left in the bottom of the bottle before you refill it from your big container. Refill the large container at a zero-waste shop.

A PLASTIC-FREE SHOWER EXPERIENCE

There is little real difference between the soap disguised within different bottles that jostle for space in our bathrooms; most likely we could substitute one for another.

Reducing bathroom clutter is more than just great advice from Marie Kondo (the Japanese organising guru), it benefits the planet at the same time.

If you can replace your facewash, liquid soap and shower gel with one —you'll save money, reduce clutter, cut your carbon footprint and slash your plastic packaging purchases. And with many soaps on the market now you'll be able to find one that suits your skin.

Additional facts

Liquid soaps require five times more energy for raw material production and nearly twenty times more energy for packaging production than bars of soap.

People typically use more than six times the amount of liquid soap (by weight) than bar soap.

Many people don't realise that wet wipes aren't just moistened tissue or woven cloths. Wet wipes are made from natural fibres, such as cotton, but also contain synthetic or plastic fibres (such as polyester and polypropylene), to which cleansers, moisturisers, detergents and preservatives are added.

WET WIPE

In today's world, there's a wet wipe designed for every possible job, so there's no more need for a licked handkerchief or a wet flannel. Or is there?

Since wet wipes first came onto the market in the US in the late 1950s, they have grown into a global market. In a world 'always on the go', convenience counts and they are now ubiquitous and there's a wipe for every situation – baby wipes, face wipes, travel wipes, toilet-training hand wipes, make-up removing wipes, antiseptic wipes, hand wipes, toilet wipes, and even floor and bathroom cleaning wipes.

Although there has been some backlash against the use of wet wipes more recently (as people discovered that wipes contain plastic), it's hard to remember a time when wipes weren't the go-to product for our increasingly mobile lives. Wipe manufacturers continue to develop ever more childcare products and wipes with ever more niche uses – face wipes with added moisturiser and toilet wipes that clean and disinfect, to name but two. And heightened concerns about hygiene have some people reaching for an antiseptic wipe for toilet seats, bins and hands.

THE IMPACTS

The huge surge in their popularity has led to the modern phenomenon of 'fatbergs' – nasty, subterranean masses that block sewers. Fatbergs are essentially a mix of wipes, congealed fat and other waste found in sewers. Unlike toilet paper, wet wipes don't biodegrade when flushed down the toilet – but it seems many people think that's how to dispose of them. Since they are held together with plastic, they don't break apart much either en route through the sewage system, which is designed for organic matter. When they get stuck, they create a blockage.

In fact, wet wipes make up 93 per cent of the material causing blockages in UK sewers. And this pattern is seen elsewhere in the world as well. Wet wipes and other sanitary products cause more than 500 sewage blockages every month in Ireland. And across the pond in 2017, a fatberg in a sewer in Baltimore, USA, caused the release of more than 4.5 million litres of sewage into a Maryland stream.

In 2017, a 250-metre-long fatberg formed of 150 tonnes of grease and wet wipes was discovered under Whitechapel in London, UK, blocking the sewage system. For a better idea of just how big this fatberg was – it was the length of two football pitches and the weight of eleven double decker buses.

If wipes make it into rivers or the ocean they cause even more problems – they can be ingested whole by wildlife or break down into smaller pieces of plastic that eventually make their way into the food chain (see also page 134). They are frequently washed up on beaches and along the shores of estuaries and rivers, causing litter and spoiling picturesque views and walks. During the Marine Conservation Society's annual beach clean in 2018, volunteers found on average twelve wet wipes per 100 metres of beach – a 300 per cent increase from just ten years earlier.

THE FIRST WET WIPES

American Arthur Julius is thought to be the inventor of the wet wipe, or what he called the 'wet-nap', in 1957 in Manhattan, New York City. Julius converted a machine designed to divide soup into portions to instead add soap to disposable towels. He believed that wet wipes were the optimal delivery system for cleaning and hygiene – better than soap and water and a towel – and that there was a real need to clean hands before and after eating in restaurants. His first big client in 1963 was Colonel Sanders who bought the wet-naps for use in his KFC restaurants.

WHAT YOU CAN DO

- Use a reusable cotton cloth or natural sponge and some warm soapy water.
- Grab an old-fashioned cotton face cloth and a bar of facial soap – the perfect substitute for make-up wipes; that's kinder to your skin, longer lasting and cheaper than a packet of wipes.
- Use hand sanitiser to clean hands when travelling or at festivals where running water is not to hand.
- Go old school. Most homes have a ready supply of warm water, detergent and a cloth or mop, all of which can be used over and over again. Disposable cleaning wipes are easy to replace.
- Moisten toilet paper with water from the tap and soap if necessary. Moist toilet wipes are a real environmental problem as, unsurprisingly, people assume wrongly they can be flushed down the toilet.
- Try washable and reusable 'wipes'. You store these cotton and bamboo cloths in a reusable plastic box or bag once moistened. The cloths come with a 'mucky' box or bag to contain the dirty wipes until you can refresh them in a washing machine.
- As a last resort, look for wipes made from cotton and those that are 'plastic free'. These wipes are made from natural fibres, water and plant extracts , but they are still disposable, block drains if flushed and cause unnecessary waste. Always bin wipes.

If you change your razor blades as recommended – every five to ten shaves – and you shave daily, you could find that you are throwing away as many as fifty-two disposable razors or replacement heads a year. Even if you eke out your razor blade to last a month, that's still twelve blades or disposable razors a year, which along with their packaging adds to the mountain of waste disposed of annually.

DISPOSABLE RAZOR

Beards and moustaches may well go in and out of fashion as does amounts of visible body hair, but at some stage or other in life most people will want to use a razor to shave their face, legs or underarms.

Our fascination with body hair, or its absence, goes back to prehistoric times when we used clam shells, shark's teeth and flints as razors. The first metal razors were made of copper and gold and date back to the 6th century BC in Egypt. The scary-looking stainless-steel straight-blade razor wielded in westerns was gradually replaced by what is called a 'hoe-shaped safety razor' from the 1880s, when King Camp Gillette invented a double-edged replaceable blade at the start of the 20th century. One of the first entirely disposable razors was invented in 1963 by American entertainer and inventor Paul Winchell. Disposable razors pitched convenience above the quality of the shave – with a focus on lower cost than razors with replaceable heads – to attract customers.

THE IMPACTS

Disposable razors are made from a mix of materials including plastic, rubber and metal as well as the various additives used in lubricating strips. Such mixed materials makes recycling very difficult and they are not accepted in household recycling bins. Add to that their packaging and the waste soon mounts up.

Razors with plastic and metal shafts and replaceable blades produce less waste than disposable razors, as you don't throw away the whole razor when the blades are spent. However, the replaceable heads, which are made from metal blades embedded in a plastic frame, are not recyclable and tend to be overpackaged.

The plastic in disposable razors can last for centuries and breaks up into microplastics.

WHAT YOU CAN DO

- Get a razor for life, then all you have to do is change the blades (actually the blade and not the whole head). They are more expensive to buy initially but way less expensive over their lifetime, as replacement blades are very inexpensive. Responding to consumer demand to move away from disposable plastic razors, steel razors are now being produced in finishes such as rose gold and brushed steel to appeal to men and women.
- Shave plastic free – some companies sell 'shaving kits' with a steel safety razor, a wooden shaving brush and a bar of vegan shaving foam – to eliminate the waste from the foam as well as the razor.
- Choose the razor with replaceable heads if limited in choice between a disposable razor and a razor with replaceable heads – as this will reduce the amount of waste produced.
- Recycle. Although you cannot recycle disposable razors and razor blades in your household rubbish, innovative waste organisation TerraCycle has joined forces with Gillette to offer a recycling service for any brand of disposable razors and plastic razors with disposable heads. Look online for a collection point near you.
- Consider a subscription service for shaving, though not all subscription services are created equal. Some provide you with a regular supply of replaceable but ultimately disposable heads (made from a plastic moulding and the blades) for a plastic or steel razor. Others, such as the brand Supply, provide a stainless-steel razor for life, replacement blades and even plastic-free shaving cream and post-shave balm.
- Disposable razors are a poor choice generally in terms of the quality of the shave and their environmental impact. But if you must have one, choose a razor with a handle made from recycled plastic, such as yoghurt pots, and designed to be recyclable.

A CLOSE SHAVE

Shaving foam and gel tends to come in an aerosol container. These containers can be recycled when they are completely empty, as can the plastic lids, but there are options that come in a lot less packaging and can even make shaving more pleasurable. For example, you could try using a shaving oil, which usually comes in glass bottles.

Shaving cream and soaps are other options. These work in much the same way – lather up with water and a brush – although some people find that shaving cream produces a good lather more easily than the soap versions. The shaving cream or soap should last for ages, unlike an aerosol, as will the brush and the packaging is minimal – a recyclable plastic container or a metal tin.

Additional facts

In the early 1990s the US's Environmental Protection Agency estimated that Americans used 2 billion disposable razors a year; in 2018, 163 million US consumers used disposable razors.

Did you know that women's razors and replacement blades are often more expensive than men's – even though they are essentially the same thing? This so-called 'pink tax' has caused controversy over the years and is in the range of a 6 per cent price differential for disposable razors in the UK.

About 600 million aerosol cans are used in the UK every year, equivalent to ten per person, and the products most commonly packed in aerosols are deodorants, antiperspirants and body sprays.

DEODORANT

Have you ever stopped to check your armpits for body odour before an important meeting or a date? Concerns about bodily smells and perspiration have not always been such a concern. In fact, in order to sell deodorants, early manufacturers had to start by convincing people that the natural act of sweating was embarrassing.

Deodorants and antiperspirants are different but are often grouped as one, as I'll do here. A deodorant, for example, kills odour-producing bacteria while an antiperspirant actually prevents sweating, rather than covering up the smell. The first deodorant, called Mum, was invented in 1888 in the US and the first antiperspirant was called Everdry and went on sale in 1903. Before that people used removable 'dress shields' in the armpits of their clothes to reduce the need for washing the whole garment.

Deodorants typically comes as a spray in an aerosol or as a roll-on. In Europe, half of all people use an aerosol deodorant while 27 per cent use roll-ons and 13 per cent use sticks.

THE IMPACTS

Aerosol deodorant sprays fell out of favour with conscious consumers in the 1980s when it was discovered that the propellant gases they contained were contributing to the hole in the ozone layer. In some effective, multilateral policy-making, the gases were banned and alternatives were found to enable aerosols to continue to be in everyday use today.

Aerosol cans contribute to the demand for aluminium (what the can is made from) and contain chemicals called propellants to help the deodorant ingredients spray out of the can. The latest innovation in deodorants is the new compressed aerosol cans – they use less aluminium, less propellant and are smaller to ship, all reducing their carbon footprint. Roll-on deodorants tend to come in a plastic container with a plastic rollerball. The container and the ball are made of different types of plastic, which unfortunately usually renders the whole item unrecyclable.

The deodorants and perfumes we use every day are contributing to indoor air pollution, which damages human health. In fact 40 per cent of the chemicals added to products like deodorant end up in the air we breathe. Spray deodorants release volatile organic compounds (VOCs, see also page 168) that interact with other chemicals in the air to form ozone and particulate matter (the tiny particles that can penetrate deep into the lungs) – both of which are thought to be harmful air pollutants.

Deodorants contain myriad ingredients, some of which have been the cause of concern among consumers. The aluminium component – used to temporarily plug our sweat glands – is hazardous in high doses, but there is no proof that its application as an antiperspirant is harmful. Other noteworthy ingredients include phthalates, parabens and triclosan (see pages 26, 40–43 and 121); make your own decision on what you would like to use on your body.

WHAT YOU CAN DO

- Use a roll-on or stick deodorant or antiperspirant instead of an aerosol spray – this will reduce the carbon footprint associated with the product. Less propellant and less metal also means less energy use and fewer emissions.
- Seek out natural deodorants, such as solid sticks in cardboard push-up sleeves and deodorant paste in glass jars that you rub into your armpits; mostly these are available in plastic-free packaging, too. You can also get deodorant powder in a recyclable plastic or glass bottle.
- Look for deodorants that are free from chemicals such as phthalates and parabens. Natural deodorants are made from bicarbonate of soda or eucalyptus to neutralise odours, coconut oil and shea oil to moisturise, arrowroot powder to absorb moisture and essential oils for fragrance.
- Use a rock. Crystal deodorants are an option. They are made of the mineral salt called potassium alum – so they do contain aluminium.
- Opt for compressed aerosol cans if you can't bear to switch from an aerosol.
- Recycle wisely. Empty aerosols can be recycled with aluminium cans only if they are completely empty, otherwise there's the risk of explosions during the sorting and recycling process.

Additional facts

If 1 million people switched to compressed aerosols (from standard versions) it would save 696 metric tonnes of CO_2 (equivalent to electricity used in 121 homes in one year) and enough aluminium to make 20,000 bikes.

In the UK 79 per cent of people buy deodorant as part of their weekly shopping. That's around 50 million people buying aerosols and plastic containers, most of which are thrown away – because four out of every ten people in the UK don't recycle bathroom items.

It is estimated that 14,000 tonnes of suncream, some containing as much as 10 per cent oxybenzone, ultimately finds its way into coral reef areas annually and is a possible contributor to coral reef bleaching.

SUNCREAM

The Sun's ultraviolet (UV) rays can damage even the deepest layers of our skin, so it's important to wear suncream. But have you ever noticed the marks that suncream leaves behind on your clothes and your car and wondered what on Earth you are rubbing on to your skin?

When the hole in the ozone layer was discovered in the 1970s, reduced levels of ozone meant more UV light was penetrating the atmosphere and reaching the Earth's surface. The more damage there is to the ozone layer, the more UVB rays reach our bodies and the greater the risk of sunburn and skin cancer.

Whether you prefer a gel, mousse, cream or oil, the active ingredients in suncream are either mineral or chemical. Mineral-based sun protection deflects or scatters UV radiation (sometimes called sunblock), while chemical-based suncream absorbs UV rays and transforms them into less damaging radiation (sunscreen).

The earliest suncreams were minerals such as calcite and clay that reflected the Sun's ultraviolet (UV) rays. It wasn't until the late 1920s that chemists caught up with synthetic sun protection, made from benzyl salicylate and benzyl cinnamate. Eugène Schueller, who founded the cosmetics company L'Oréal, was a keen sailor with pale skin. He charged his chemists with inventing a cream to prevent sunburn. In April 1935 they presented him with a sun oil, later called Ambre Solaire, that contained benzyl salicylate to filter out UVB rays, the type that causes the skin to burn.

Skin cancer, specifically melanoma of the skin, is now the nineteenth most common cancer worldwide. Australia had the highest rate of melanoma in 2018, followed by New Zealand. An estimated 86 per cent of melanoma skin cancers in the UK (around 13,600 cases) every year are linked to too much exposure to sunlight and sunbed use, yet in many countries people are still underusing suncream.

THE IMPACTS

Suncream ingredients have been shown to damage coral, accumulate in fish and the environment and disrupt hormones in fish and amphibians. Studies have identified the chemicals that filter out the UV filters (such as oxybenzone, octocrylene, octinoxate and ethylhexyl salicylate) in almost all water sources around the world. The filters are not easily removed by common wastewater treatment techniques, so when you shower suncream is washed down the drain and out into the environment. As a result, such chemicals have been discovered in various species of fish worldwide (for instance, in cod liver and white fish in Norway and in white fish and mussels in Spain), which has potential consequences along the food chain right up to the level of us.

It is clear that suncream plays an important role in preventing skin cancer and it might seem that more natural suncreams would be a good choice – but it's important to know that they work. There is no legal standard for the term 'natural' as it applies to personal care products. Analysis of four years of suncream testing data for consumer reports found that only 26 per cent of natural suncream lived up to their SPF claims, compared with 58 per cent of chemical sunscreens. Also, many US suncreams are not sold in the EU as they do not meet its more stringent requirements.

THE FUTURE OF SUNSCREEN?

Inspired by coral's ability to protect itself from harmful UV rays, Australian scientists are working on a new sunscreen that would mimic coral's natural sun protection. By trying to recreate in a lab the protective compounds corals produce, this new type of sunscreen could protect us and be safe for coral. Win, win.

WHAT YOU CAN DO

- Choose mineral-based sunblocks that use titanium dioxide and zinc oxide – they are safer to use and better for the environment than chemical-based sunscreens containing oxybenzone. Some mineral suncreams contain the minerals as nanoparticles to make them easier to apply. Choose suncream that is certified as reef safe.
- Look for suncream with the 'Protect Land + Sea' certification seal of approval, which means that the product has been laboratory-tested to verify that the product is free of known pollutants of different environments.
- Recycle empty plastic suncream bottles if fully clean and dry.
- Reduce your exposure to the sun by covering up with clothing, hats and sunglasses and avoid being out in the intense sun during the peak hours of 10 a.m. to 4 p.m.
- Wear sun-protective clothing rather than suncream when snorkelling or swimming on reefs to protect the coral.

THE PLASTIC CONUNDRUM

When the BBC documentary series *Blue Planet II* aired in 2018, the public responded in shock to the images of plastic in the ocean, entangling wildlife and even being eaten by seabirds and other marine life, causing them to suffer and die.

Despite its recent reputation as a cause of ocean pollution, plastic is not intrinsically bad. Used in the right way it remains a wonder material – we need it for medical devices and prosthetic limbs as well as to safely contain toxic chemicals. It is a question of using plastic appropriately.

AN OCEAN-GOING ISLAND OF PLASTIC

The Great Pacific Garbage Patch, discovered in 1997 between Hawaii and California, is the largest accumulation of marine plastic in the world. It is huge and is one of five places in the world where ocean currents cause plastic waste to accumulate in a gyre; the others are in the South Pacific, the North and South Atlantic and the Indian Ocean. The Great Pacific Garbage Patch covers an areas three times the size of France and contains an estimated 1.8 trillion pieces of plastic weighing 80,000 metric tonnes.

Plastic is a menace when it becomes litter and marine pollution. Marine mammals get caught up in plastic fishing nets, ropes and beer can rings, for example, and turtles have been known to eat plastic straws. It has been estimated that 40 per cent of marine mammals and 44 per cent of seabirds have ingested plastic, mostly as tiny microplastics. Annual deaths due to ingested plastic number up to 1 million sea birds and 100,000 sea mammals.

PLASTICS AND CLIMATE CHANGE

Another major problem is that plastic is a major contributor to climate change. Most plastics are made from oil and natural gas, both fossil fuels, and greenhouse gases are emitted throughout the life cycle of plastic, from the extraction of the raw materials and the making of the plastic right through to when it is recycled or becomes waste. It is estimated that by 2050, the production and incineration of plastic could be responsible for 2.75 billion metric tonnes of CO2 globally, equivalent to the annual emissions of 706 coal-fired powered plants and enough to put the world's chances of staying below 1.5°C of global warming out of reach (see page 10). Even in 2019, the production and incineration of plastic was already responsible for more than 850 million metric tonnes of greenhouse gases— equivalent to the output of 189 coal-powered plants.

Plastic does not biodegrade. In a landfill, plastic just sits there while the organic material around it decomposes. But in the ocean, it's another story. The UV rays from the sun and the attrition of the water break plastics down into ever-smaller pieces – micro- and nanoplastics. These tiny pieces can be ingested by marine life, causing them harm and ultimately entering our food chain.

These nano- and microplastics not only pose a danger because of their size, they also contain the chemicals and additives used to make the plastic, which are toxic to marine life and to humans. And to make matters worse, the plastic particles attract other impurities, thereby concentrating the toxic effect. A study conducted at Hawaii University, and published in 2018, found that as plastics decompose they also emit traces of methane and ethylene, two powerful greenhouse gases. The emissions occur when plastic materials are exposed to solar radiation in water or in the air, with much higher emission rates in air. This means that plastics represent a source of greenhouse gases that was little understood until recently.

SINGLE-USE PLASTICS

The main problem with plastics is that we use them to make objects that very often don't need the durability and long life that the material provides. So-called single-use items – coffee cups, carrier bags, packaging, disposable forks, cotton buds, stirrers and straws – may be used for typically 20 minutes; the plastic they are made from can last for centuries. Such disposable items are thrown out after one use and end up as litter.

All kinds of single-use plastics are now on the way out; as far back as 2002 Ireland led the way with the introduction of a levy on plastic bags, which led to a 90 per cent reduction in the use of disposable plastic bags. Antigua and Barbuda and Kenya followed suit, banning plastic bags outright in 2016 and 2017, making it illegal to even bring a plastic bag into the country. New Zealand outlawed plastic bags in 2019, while Australia, the US and Canada are yet to impose a ban. Barbados banned single-use plastics from plastic bags to takeaway containers in 2018 and EU countries and the UK will ban ten single-use items by 2021.

WHAT'S NEXT FOR PLASTIC?

Throughout the book you'll find people using innovative ideas and novel technologies to create reusable or biodegradable versions of single-use items – from keep cups and paper cotton buds to edible straws and refillable packaging – as well as other plastic-heavy products. The future of plastic might lie in plant-based plastic, made from renewable resources such as sugar cane, seaweed and cassava. Plastics need to evolve to be reusable, compostable or recyclable with an emphasis on avoidance where possible – so that we use less plastic packaging and produce less cheap and throwaway plastic 'stuff'.

COULD PLASTIC-EATING BACTERIA BE THE SOLUTION?

Good news! Scientists have discovered a strain of bacteria that 'eats' the plastic in PET bottles helping them to decompose. This is a hopeful development for plastic in landfills but not a replacement for wiser use of plastic. A concern is if the plastic-eating bacteria were loose in the environment, they could damage plastics that we need to be durable and strong, such as plastic barrels holding chemicals.

Research from Garnier found that over 56 per cent of people in the UK (4.5 million people) don't recycle bathroom products because of the inconvenience of rinsing it out and putting it in the recycling bin. A Unilever study in the US revealed similar habits and attitudes. Only 34 per cent of Americans recycle bathroom packaging, claiming confusion (42 per cent), lack of certainty that the product would actually get recycled (27 per cent) or laziness (20 per cent).

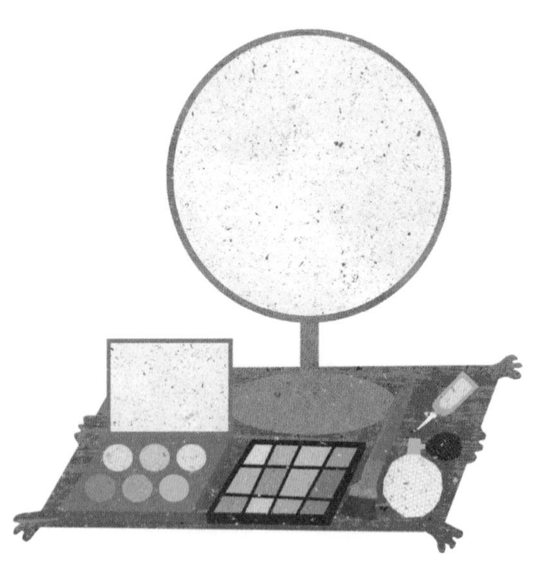

COSMETICS

Some of the key ingredients in make-up remain the same as they did in Cleopatra's time: minerals, plants and oils. But modern cosmetics include other ingredients as well – preservatives, solvents and plasticisers (used to make things soft and flexible). In 2015, a US NGO found that the average US woman uses twelve personal care products and/or cosmetics a day, containing 168 different chemicals.

The Egyptians were some of the first people to use make-up to improve their looks, around 3000BC, as distinct from body paint that was used by early man for ceremonial or camouflage purposes. Ancient paintings reveal that men and women used dramatic make-up made from minerals, plants and fats to highlight their eyes, lips and cheeks. Cleopatra, for example, is always depicted with dark kohl under her eyes, and minerals such as malachite, pyrite and red ochre were used by her and others to add colour to their faces.

Much later on, in 16th century Elizabethan England, make-up was used to differentiate the aristocracy from the lower classes. Although the mixture of lead and vinegar, called Venetian ceruse, whitened the skin and smoothed the complexion, it was also toxic and caused skin discoloration, hair loss and rotten teeth.

THE IMPACTS

While the average US woman may be exposed to personal care products containing 168 different chemicals every day, men may use fewer products (about six), but they are still exposed to about eighty-five such chemicals daily; while teens, who use an average of seventeen personal care products a day, are exposed to even more.

When it comes to the environment, the cosmetics industry is a major contributor to waste. More than 120 billion units of packaging are produced every year by the global cosmetics industry and most of that is not recyclable. Even when packaging is recyclable, it is very often not being recycled (see opposite page).

On top of the waste from packaging, consumers also waste the actual cosmetics. The same study reveals that men are more sustainable – they own twelve products on average, and use them all. Overall it seems that women have a few favourite products and they may try others once or twice but then leave them aside, untouched and unused.

How good are you at buying only what you need? Look in your bathroom cabinet, that drawer in your bedroom, your handbag and see what is lurking there. An eyeshadow set with only two colours used? Four lip glosses but only one that you really wear? Even the way companies market their make-up creates waste – think of the offers to buy two face creams and get a free bag of mini-pots of other products. Ask yourself first, would you actually use them?

Add to this the fact that it can be hard to get empty or out of date make-up containers clean and that many containers are a mix of plastic, glass and metal, and it may feel almost inevitable that all these things end up in landfill and incineration.

Parabens are added to make-up as a preservative to prolong shelf-life. A US study in 2011 found parabens in 66–87 per cent of women's cosmetics and 77–82 per cent of lipsticks and lip liners, as well as in shampoos, sunscreens and even wet wipes.

Parabens are thought to interfere with reproductive hormones in people (the effects are seen in aquatic life, too), leading the European Commission to take the precaution of banning five parabens from cosmetics in 2015.

MICROBEADS

Apparently, one tub of facial scrub can contain as many as 300,000 microbeads. These tiny pieces of plastic are used in facial scrubs to exfoliate skin and are also found in some bubble baths, shower gels and toothpaste. They are so small that they bypass the wastewater treatment and end up in waterways where they are mistaken for food by aquatic animals and get into the food chain.

Countries including the US, New Zealand and the UK have banned microbeads in cosmetics; a ban is planned in Ireland and there is a voluntary ban in place in Australia.

If you want to know if there are microbeads in your facial scrub, look out for these ingredients – polyethylene terephthalate (PET), polypropylene (PP), polyethylene (PE), and polymethyl methacrylate (PMMA).

Another group of chemicals called phthalates (see also page 26) are used to make nail varnish, gel nails, hairspray and some perfumes. Phthalates are also thought to be damaging to our health and, as a result, seven of them are forbidden as ingredients in cosmetics in the EU, but are still permitted in the US and other jurisdictions.

The recent rise in vegetarianism and veganism has led to demand for plant-based cosmetics (without dairy or honey, for example). This has grown out of a much longer-running campaign to end animal testing of cosmetics, first championed in the 1970s by Anita Roddick and the Body Shop. The EU, India, Norway, South Korea, New Zealand and most recently Australia have all banned animal testing of cosmetics, but it still takes place in the US and China, where it will be phased out by 2020.

Overall, it can be hard to tell what 'organic' or 'natural' mean on cosmetics labelling. You can look out for certifications (see pages 62–63) but be careful as some are country specific and have limited independent oversight.

WHAT YOU CAN DO

- Finish your make-up before you buy more. Cut open old tubes and rinse out bottles to reduce waste. Empty containers and clean them for recycling where possible.
- Resist buying multipacks and 'buy one get one free' offers, unless you are sure you will use the products. Buying only what you really need and know you will use can significantly reduce waste.
- Pass it on. If you buy make-up and find it's not for you, give it away to a friend or family member promptly so that the product doesn't go out of date.
- Research before you buy. Look out for brands with a reputation for making natural products – such as the Body Shop and Lush.
- Look for brands that refill or accept empty containers for recycling.
- Choose water-based, solvent-free nail varnish with natural pigments and oil. It will be better for you and better for the environment. And as for false nails, really you just have to go without these colourful plastic accessories for the benefit of the planet!
- Recycle your old cosmetics containers – recycling firm TerraCycle have partnered with Garnier to create a recycling programme for beauty products in the US and Australia. It accepts all kinds of haircare, skincare and cosmetics packaging. L'Occitane and Burts Bees also run recycling programmes with TerraCycle.
- Only buy cruelty-free products. This is easier in the EU where animal testing is not permitted.
- Go without make-up altogether or have some make-up-free days. Let your skin breathe and use less product.
- Turn down free promotional samples as the small containers create waste and they are likely to remain unused.

MANICURE ALERT!

If you love nails, then look again. Most nail varnish is made of plastic, as are false nails. As nail varnish chips off, it becomes a microplastic pollutant.

Additional fact

A 2014 study by Vaseline in the UK found that for every ten cosmetics a woman buys, she only uses one! This equates to the average woman wasting 5,846 beauty products in her lifetime, amounting to £180,000.

SUSTAINABLE COSMETICS

With consumers shunning chemicals and plastics, the cosmetics industry is responding to a growing demand for ethical, natural and sustainable products with a range of minimal, refillable or recyclable packaging.

The Body Shop was first to market by a long way, bringing cruelty-free and responsibly sourced cosmetics to the high street in the late 1970s. Since then, brands are innovating to reduce the impact of their packaging and increase the benefits of their cosmetics to the planet. Zero waste is the ultimate goal.

The UK company Neal's Yard Remedies sets the bar high in terms of their sustainable credentials. In 2014, it was awarded 100 out of 100 for ethics by the Ethical Company Organisation, which operates across the EU. It sources its ingredients from small-scale growers who use organic, sustainable and wild-harvesting practices. As well as being carbon neutral (by installing solar panels and by offsetting their emissions by supporting a forest protection project in Madagascar), it pledges to make all of its plastic bottles from 100 per cent recycled material by 2025.

Other companies setting a great example when it comes to packaging include:

- Lush leads the way bringing naked or zero-packaging soap, shampoo, conditioner, shower gel, face wash and body lotion to the high street.
- Kjaer Weis sell their organic make-up in refillable metal containers – everything from lipstick and cream blush to compacts and mascara. The container lasts for life and you buy refills when you need them. The refills come in cardboard packaging that are easily recycled.
- Kilian perfume is described as 'eco-luxe'. Each fancy bottle of perfume is designed to be kept for life and refilled via a simple, recyclable glass bottle.

The UK uses more plastic cotton buds than any other European nation – a staggering 13.2 billion of them each year.

Cotton buds with plastic spindles will be banned in the EU by 2021.

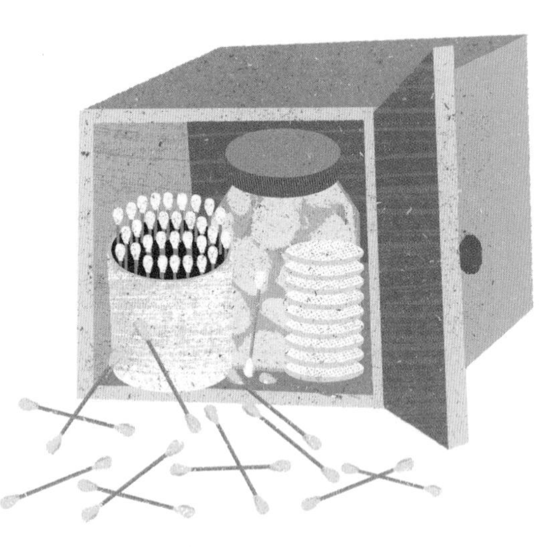

COTTON BUD

Whether they're used for removing eye make-up or for the strangely pleasurable sensation of cleaning out ear wax, cotton buds are to be found lurking in almost every bathroom cabinet. But who knew such a mundane cleaning item could escape into the marine world? Anyone who saw the seahorse with its tail wrapped around a cotton bud as part of the footage on *Blue Planet II* now knows all too well about the polluting nature of this object.

Cotton buds were invented by a Polish-born American man called Leo Gerstenzang after he watched his wife use a toothpick and a piece of cotton as a makeshift tool to clear their baby's ear. He called his original invention the Baby Gay but renamed it a Q-tip in 1926 (where the Q stood for quality); Q-tip remains a household name in many areas of the world.

Cotton buds are made of three simple parts – the stick or spindle, which used to be made of wood and is now made of paper or plastic; the absorbent end made of a swab of cotton wrapped around itself; and the packaging, which can be cardboard or plastic or both.

THE IMPACTS

The main problem with cotton buds arises when they are flushed down the toilet, which it seems a surprising number of us do. A survey in Ireland in 2015 found that 26 per cent of people flush used cotton buds down the toilet. The plastic sticks that make up the spindles are small enough to pass through sewage treatment plants and ultimately enter rivers and the sea where they become plastic pollution.

According to the Switch The Stick campaign in the UK, plastic cotton bud sticks comprise over 60 per cent of sewage-related beach litter. A marine pollution study on the Tyrrhenian

Coast in Italy published in 2016 found that plastic cotton bud sticks made up more than 30 per cent of all the litter collected. Most of the cotton buds originated in domestic sewage and were washed into the sea from rivers after passing through sewage treatment plants.

When cotton buds are eaten by marine creatures they can get stuck in an animal's stomach, meaning the animal feels full, doesn't eat and will eventually starve to death. Also as the plastic spindles break down into microplastics in the sea, they absorb other toxins present in the sea, so pose a double risk of poisoning marine life.

In terms of their manufacturing, cotton buds made from plastic consume oil as a raw material as well as additives and chemicals to give the plastic colour and flexibility. Since paper sticks are made from trees, check that the manufacturer sources recycled and FSC-certified paper – you should see a label to this end on the packaging.

Cotton itself requires a lot of water to grow and many farms use vast quantities of pesticides and fertiliser in its production, leading to water pollution and soil degradation (see also pages 105–107).

WHAT ABOUT COTTON WOOL AND COTTON PADS?

Cotton wool and cotton pads are single use, disposable items. They are convenient and, in some cases, important for hygiene and health and safety, but if your main use for them is wiping a baby's bottom or taking off your make-up there are alternatives.

Cotton is a very polluting commodity – using lots of water, pesticides and herbicides – with negative impacts on biodiversity and human health. So, if you do need cotton wool look for brands that are certified as 100 per cent organic.

Look for reusable cotton pads made from organic cotton or bamboo. They are washable and come in various sizes suited to different tasks, such as removing eye make-up. And while a pack of ten such pads costs more than a packet of cotton pads, because you only have to buy them once they will save you money in the long run. Using reusable cotton pads will reduce your bathroom waste dramatically, while adding only a little to your laundry pile.

WHAT YOU CAN DO

- Give them up. Doctors advise never to put anything bigger than a finger in our ears to avoid infection. Ears are designed to be self-cleaning, apparently.
- Never flush cotton buds down the toilet. Throw them in the bin.
- Choose cotton buds with paper or bamboo spindles over plastic ones. Check the package but most of these can be composted after use.
- Opt for cotton buds with minimal packaging that can be recycled.
- Switch to a reusable, washable cotton pad for touching up or removing eye make-up if that's the only time you use cotton buds.
- Cut down on the number of cotton buds you use, if you use them to apply or tidy up make-up. And switch to those with sticks made from paper or bamboo, not plastic.
- Embrace alternatives. If you use cotton buds for cleaning hard to reach spots, try an old toothbrush instead or, for smaller jobs, wind a piece of tissue around a toothpick.

Medical waste, including swabs, bandages and adhesive plasters, is among the ten most common marine litter items in Europe. Most sea swimmers and beach walkers can attest to seeing used plasters floating in the water or washed up on the beach.

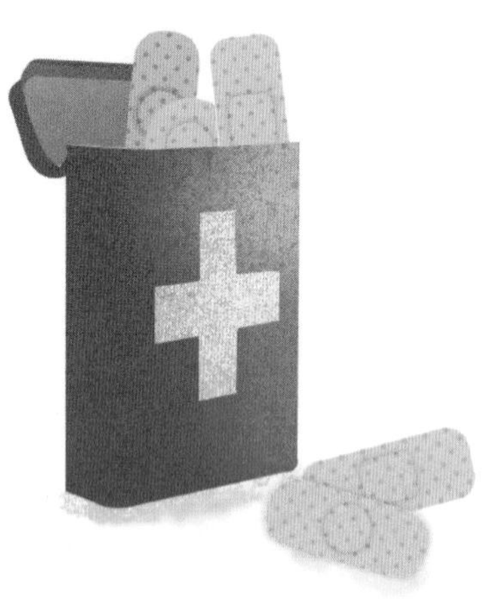

PLASTERS

Earle Dickson, an employee working at Johnson & Johnson in the US, invented the first self-adhesive bandage for his accident-prone wife in 1921. After some refining Dickson patented the BAND-AID in 1926: a thin strip of adhesive and gauze covered by a protective layer that you remove to apply it. Not longer after in 1928, Elastoplast dressings were devised by employees at TJ Smith & Nephew, a firm with origins as a small dispensing chemist in Hull, UK.

Adhesive bandages or plasters are made from plastic or woven fabric, the backing is either a coated paper or plastic and the adhesive is commonly an acrylate, which is a petroleum-based thermoplastic. The absorbent pad is often made of cotton, and there is sometimes a thin, porous-plastic coating over the pad to keep it from sticking to the wound. As a result, all conventional plasters contain plastic, which means they are not biodegradable.

Nowadays there's an array of plasters available: plasters for fingers, antibacterial plasters, kids' plasters, plasters for sensitive skin and even spray-on plasters, which come in an aerosol can. Spray-on plasters may be convenient but these plasters are still made from plastic and the residue should be disposed of carefully.

THE IMPACTS

How often have you seen a plaster float by in the swimming pool or seen one stuck in the sand on the beach? They fall off easily while we swim or exercise, often without us noticing. Once in the water or on the beach, they become litter and marine pollution, posing a threat to wildlife.

> ### SINGLE-USE BAMBOO PLASTERS
>
> Patch strips are made by the company Nutricare in Australia and are a single-use alternative to plastic and latex plasters. They are made entirely from bamboo, which is naturally antibacterial and breathable, from the gauze to the adhesive. They are certified as sterile and suitable for medical use. The plasters are fully compostable and decompose within ten weeks.
>
> Patch strips use aloe vera, charcoal and coconut oil as antiseptics and to help wounds heal, instead of chemicals, such as thimerosal, which are commonly found in plasters and can cause skin reactions in some people.

Plasters don't break down; they are single-use plastic items that can't be reused or recycled as they are usually contaminated with blood. They are made from various types of plastic, all of which are derived from oil and may contain additives such as dioxins and phthalates (see pages 26 and 137), which are thought to be harmful to the environment.

Some plasters are made from latex or natural rubber, so they come from a renewable resource (rather than the oil that plastic is made from), but they still take a very long time to biodegrade and they aren't suitable for people with latex allergies.

Reusable bandages are the most sustainable solution but they are not always practical or hygienic. New innovations include plasters made from copper and a tiny generator that uses an electrical current to helps wounds heal more quickly. The plasters are powered by your body's movements and the lightweight device is small enough to be worn like a regular plaster. Hopefully they can be designed to be reusable too! We'll watch this space.

WHAT YOU CAN DO

- Dispose of plastic plasters carefully. Make sure they go in the rubbish bin and don't end up floating in the sea, as they are not compostable or recyclable.
- Look for plasters made from natural materials, such as bamboo, coconut oil, charcoal and aloe vera. Australian firm Nutricare makes compostable bamboo plasters with charcoal, aloe vera and coconut oil to help healing (see box, above).
- Wrap wounds in a sterile organic cotton fabric bandage to avoid waste completely; you can wash it after use or compost it if it is just a small piece.
- Use a plaster only if you really need one – for grazes and small cuts simply use a smear of antibacterial ointment to protect it while allowing the skin to heal. If you have a deep wound, however, there may be no alternative to a plaster to keep it clean.
- Help clean up beaches. Used plasters are a frequent and unwelcome discovery by beach cleaners. Next time you are at the coast do a 2-minute beach clean and see what you find.

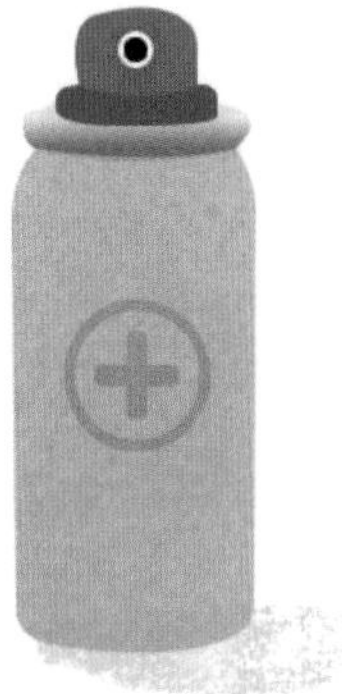

The Marine Conservation Society in the UK reported an increase in sewage-related debris, including sanitary towels, tampons and applicators in its 2017 nationwide beach clean. They found an average of twenty-three sanitary pads and nine tampon applicators per kilometre of British coastline.

SANITARY PRODUCTS

It is hard for any modern woman or girl in a developed country to imagine life before sanitary products. Being able to work and attend school while menstruating has been revolutionary for millions around the world. Before specific products were invented, women and girls used grass, leaves, wool, cotton and fur or moved to menstruation huts for the duration of their period.

The first sanitary towels were used when French nurses tending injured soldiers in wars in the late 1880s repurposed absorbent, wood-pulp bandages for their own needs. These creations inspired the first commercially available sanitary towels that had loops on each end and were held in place by a belt. However, they were expensive and didn't perform very well as they didn't stay in place. The addition of a sticky strip on the base of the pad helped and over time the design of pads has improved – they are now slimmer, more absorbent and stay put.

Tampons are a more recent invention. In 1929 an American physician – Dr Earle Haas – invented the applicator tampon and patented it in 1931. Tampons are now available with and without applicators.

Despite these advances in developed countries, many women still rely on rags to deal with their periods, and the lack of affordable and accessible sanitary products is a major impediment to girls' education. Taxes on sanitary products have been reported to make sanitary products unaffordable for families on low incomes, with so-called period poverty reported to be an issue in some parts of the UK. A survey of more than 2,000 people in 2018 found that a quarter of Scottish students at school, college or university struggled to access sanitary products, prompting a project to provide access to free period products for all students across Scotland.

THE IMPACTS

The fact that most sanitary products are disposable means they create a lot of waste. Although reusable menstrual cups were developed shortly after disposable pads, they did not become popular, perhaps because they weren't marketed as effectively or because they are a little trickier to use. It is estimated that each menstruating woman throws away 125–150kg of menstrual products in her lifetime. When tampons or sanitary towels are flushed down the toilet they block sewers and pollute rivers and the ocean.

Sanitary towels persist in the environment – it can take 500–800 years or more for them to break down. Tampons, pads and panty liners generate more than 200,000 tonnes of waste each year and much of that is plastic, especially when you consider the individual bags and wrappers they often come in. Up to 90 per cent of a pad is made of plastic, which will ultimately be incinerated, sit for centuries in landfill or gradually break down into microplastics if it ends up in rivers and the sea.

A life cycle analysis of tampons conducted in 2006 by the Royal Institute of Technology in Stockholm found that the plastic in tampon applicators is the main source of carbon emissions in their production. So, non-applicator tampons produce less waste and fewer emissions.

WHAT YOU CAN DO

- Use non-disposable sanitary products. You can buy reusable, washable sanitary towels that come in a range of colours and designs. Silicone menstrual cups come in different sizes and are seeing a resurgence in popularity. Or try washable and reusable period underwear (see also page 146). Reusable products may cost more than a packet of tampons or towels initially but save you money in the long run as you can reuse them over and over again. A menstrual cup, for instance, can last up to ten years when cared for properly and costs £20–25, as compared with £900 for a ten-year supply of tampons.
- Never flush disposable menstrual products down the toilet – even if advertised as flushable. Sewage systems are not designed to deal with them; they block pipes and pollute the marine environment. Always put them in the bin.
- Choose tampons and pads made from organic cotton and without plastic. Conventional cotton production accounts for 16 per cent of insecticide released into the environment worldwide, and exposes growers and pickers to toxins.
- Avoid fragranced products – they are more polluting (with more chemicals) and are also more likely to cause skin irritation.

Additional facts

It is estimated that tampons are used by 100 million women globally.

Women use between 10,000 and 16,000 sanitary towels or tampons in a lifetime.

Women spend on average £90 a year on disposable sanitary products.

PERIOD KNICKERS

The invention of period knickers could literally change women's lives. They could make sanitary protection accessible to the millions of women who have to do without and reduce the waste associated with periods for all women.

These washable knickers can be used over and over again. They are designed to be absorptive and to cater for different flows – light, medium and heavy (equivalent to four tampons). You can wear them on their own or use as a back-up with a menstrual cup for heavy flow. They are thin and feel just like regular underwear. After wearing, you rinse the pants, wash them on a cold cycle with laundry detergent only – no softener – and line-dry them.

For those women used to tampons and pads, these knickers provide security against leaks and a real alternative to disposable sanitary products. For the women and girls of the world who have been making do with rags, these knickers provide a much more hygienic and effective reusable option.

The knickers cost about £30 a pair but they avoid the need to buy tampons and pads every month. If you look at the numbers, period pants work out cheaper in the end. On average, a woman uses twenty-two disposable products per period – equivalent to £4.15–4.60 per month. So, if you buy two pairs of period knickers it equates to the disposable products you would have used in the course of a year, but you can wear the knickers for years to come.

Many of the companies who make period knickers partner with organisations that provide access to menstrual hygiene products for women who cannot afford them, in both developed and developing countries. So, by supporting them you can also help other women access safe, hygienic and reusable sanitary solutions.

Globally, toilet paper production stands at about 84 million rolls per day.

TOILET PAPER AND TISSUE

We use roughly eight to nine sheets of paper per trip to the toilet and an average of fifty-seven sheets of toilet paper a day!

We have the Chinese to thank for toilet paper. It was widely available in China as far back as the 6th century, but it wasn't mass produced around the world until the late 19th century. For centuries people used water, leaves or newspaper for toilet duties – in fact, many still do today. The first commercially packaged toilet paper was sold in the US in 1857 by Joseph C. Gayetty, as loose, flat sheets of paper medicated with aloe vera.

Where wiping and blowing noses are concerned, there have always been tissue-free options – such as snorting or using the back of your hand – but these are obviously far from hygienic! The origin of paper tissues goes back to the search for an alternative to cotton bandages during World War I (when cotton was in short supply) and the invention of cellucotton from cellulose. The Kimberley-Clark company brought the idea from the battlefields back to the US from Europe and used it to develop new products, including tissues, initially targeted at women for make-up removal and later branded as Kleenex. Nowadays, tissues and toilet paper are considered an important tool in avoiding the spread of colds and flu and other diseases.

THE IMPACTS

Paper comes from trees, which absorb and store CO2, thereby playing an important role in addressing climate change. However, once trees are cut down that benefit is lost. Illegal logging and single-species forests, also known as monocultures, add to the negative impact of felling trees as they lead to habitat destruction and the loss of wildlife.

A life cycle analysis of tissues carried out by Kimberly-Clark found that the biggest

SUSTAINABLE LOO PAPER

Who Gives a Crap is an Australian company that sells environmentally friendly toilet paper made from 100 per cent recycled paper or bamboo and wrapped in paper rather than plastic. They donate 50 per cent of their profits to building toilets for the 2.3 billion people around the world who don't have one.

contribution to the environmental impact is the energy and water required for the paper-pulping process. The study also found that making tissues from bamboo has less environmental impact than making them from wood. The manufacturing of paper contributes to water pollution and the energy involved contributes significantly to the carbon footprint of the packet of tissues or roll of toilet paper, as does the transport of the product to your supermarket shelves (see also pages 60–61). Using renewable energy to power manufacturing and electrifying transportation can help to reduce the carbon footprint.

Recycled paper makes up 35 per cent of all fibres going into tissue production globally and demand for recycled paper exceeds supply. Furthermore, wood fibres can only be recycled three to five times, so a flow of new wood into the system is inevitable.

As single-use items, tissues and toilet paper contribute to waste. In a landfill they decompose to release methane, a powerful greenhouse gas, and they contribute to litter, although they do ultimately break down. The packaging, though, is another factor. Toilet rolls are mostly sold wrapped in soft plastic film that cannot be recycled, and tissues come in cardboard boxes or in individual plastic-wrapped packs.

WHAT YOU CAN DO

- Embrace the handkerchief. Buy, make or steal one from your dad or granddad. Say goodbye to the mountain of dirty tissues and protect your poor nose when you have a cold with a soft cotton hankie. Wash them at 40°C (60°C if you have the flu) and reuse over and over. They get softer the more you wash them.
- Choose toilet paper made from FSC-certified unbleached and ideally recycled paper. Look out for toilet rolls that are unwrapped or wrapped in paper as an alternative to plastic. Larger rolls also mean less waste. Choose FSC-certified, unbleached paper, ideally recycled tissues with easy-to-recycle boxes and avoid small packets of plastic-wrapped tissues if you can.
- Try bamboo toilet paper (see above). It grows more quickly than trees and is sustainable.
- Compost dirty tissues in a compost bin or put them in your brown bin.

Additional facts

In the 16th century, Europeans repurposed a cloth used as a head covering (called a kerchief) as a cloth for wiping hands, faces and noses. They called it, unsurprisingly, the handkerchief.

In countries such as the UK, Ireland, New Zealand and Australia, the sewage systems are designed to deal with toilet paper, so it is OK to flush it down the loo. Watch out, though, when you are on holiday as some countries prefer you to put toilet paper in the bin.

Between birth and being potty trained, a baby uses 4,000–6,000 disposable nappies or twenty to thirty reusable nappies.

NAPPY

As if having a baby wasn't confusing enough, parents-to-be now have to consider information and opinions on which nappies are best for their baby and best for the environment. Nappies are big business, so the people who make and sell them have a vested interest in getting you to choose theirs, by selling you convenience, cost savings and eco-credentials.

For millennia people managed without nappies, using leaves, fur, moss, linen and wool to catch the worst of it, or mothers swaddled babies tight, wrapping both baby and poo in a smelly bind. Other cultures let babies go free (and in some cases still do), whisking them outside if time allows or designing clothes with flaps. Mothers learn their babies' cues, so that they know when they need to go.

By the late 1800s in Europe, it was common to use squares of cotton or linen cloth and a safety pin as a nappy. But leaks were common and women put their minds to finding a better solution. In 1946 a woman from Indiana, Marion Donovan, made what she called a 'boater' from a shower curtain, a type of waterproof cover to put over a nappy. Meanwhile in the UK, Valerie Hunter Gordon focused on absorption and created the 'Paddi', which was a cellulose pad worn inside a plastic cover that closed with snaps.

In the 1950s the focus switched to making an all-in-one disposable and affordable nappy, but market penetration was low. At that time in the US just 1 per cent of nappies were disposable. Most mothers were using cotton terry nappies and then soaking and washing them after use. When Proctor & Gamble patented Pampers in 1961, the evolution of the disposable nappy really began.

By 1970 American babies were using 317,000 metric tonnes of disposable diapers, making up 0.3 per cent of US municipal waste, rising to 1.75 million metric tonnes of nappies, making up 1.4 per cent of municipal waste within just ten years.

It is estimated that around 3 billion disposable nappies are used every year in the UK, around 8 million per day. Nowadays parents have a choice between disposable nappies, plastic-free or biodegradable nappies or reusable cloth nappies. But which type of nappy is best for the planet?

THE IMPACTS

Disposable nappies are made of several layers of different materials: wood pulp from trees to absorb moisture; polypropylene in the top sheet to protect the baby's skin from getting wet; a polyethylene outer cover to prevent leaks; other plastic and adhesives in the tapes.

Reusable cloth nappies, on the other hand, are made from cotton and are held in place with Velcro tabs. You can use disposable or reusable liners to protect the baby's skin and to capture any poo, making it easier to dispose of.

The raw materials used to make a nappy and the manufacture of disposable nappies are the main contributors to the overall environmental impact and carbon footprint of a nappy. In fact, 60 per cent or more of their footprint relates to the water and wood used to produce the pulp, the petroleum needed to make the plastic and the energy used to produce the nappy. The cutting down of trees to make the pulp is the main contributor to the greenhouse gas emissions associated with a nappy. The second-biggest environmental impact is the end-of-life stage, in other words when a nappy becomes waste and is either sent to landfill or incinerated. The fact that nappies contain a lot of plastic means that they do not biodegrade. Sadly, they are a common source of litter in the countryside, on beaches and even in the ocean.

There are biodegradable and plastic-free nappies on the market, but they still require raw materials, such as wood pulp, and energy to be made, so the production impacts remain high. When such a nappy is disposed of it either goes to landfill where it slowly decomposes, releasing methane gas that contributes to climate change, or goes for incineration. Sad as it may be for eco-minded parents, these type of nappies really have few benefits over a regular disposable nappy largely because of their single-use nature. Composting nappies may sound like a good idea, and while it works to break down the pulp, it is not recommended due to the pathogens present in human waste. And industrial composters do not accept nappies of any type for composting.

For reusable nappies, the carbon footprint is primarily divided between growing the cotton and making the nappy, followed by the energy used to clean the nappy after use. So, washing on a lower temperature and line-drying can dramatically cut the carbon footprint of reusable

KEEPING NAPPIES OUT OF THE BIN

A specialised plant in Treviso, Italy, is the first in a series to recycle disposable nappies, up to 65 million of them a year. The plant in Italy can process 10,000 tonnes of waste each year.

The plant uses steam to sterilise and separate the nappies into three parts – cellulose, super-absorbing polymers and mixed plastics for reuse in industry.

nappies. Washing does use water and washing detergent, which adds to the environmental footprint of cloth nappies, but this is outweighed by the waste avoided and raw materials saved.

The key advantage of cloth nappies is that they are not thrown away. You only need twenty to thirty of them per baby and you can hand them down from one baby to the next. If convenience is important, there are nappy washing and rental services to reduce the load, and if the upfront cost of buying cloth nappies is an issue, you can borrow nappies from a nappy library or buy second-hand nappies. As the nappies are only used for a few months before a baby moves up a size, and because they are easier to clean nowadays than in the past, they are perfectly reusable.

Reusable nappies save you money as they cost one-third of the cost of disposables over the time your baby is in nappies. Four out of five users of cloth nappies use disposable nappies for convenience when travelling, as washing reusables may not be an option.

Reusable nappies have come a long way from the terry nappies and nappy pins of old. Nowadays, there are shaped nappies, all-in-one nappies and nappy covers. Modern versions are still made from cotton, and increasingly organic cotton, but they can also be made from bamboo and fleece. Guppy bags (see also page 108–109) can help to catch the microfibres released from fleece liners when you wash them, especially as nappies need to be washed a lot.

Nappy liners made from paper (a bit like toilet paper) are used to line the nappy and catch the poo and can be flushed down the toilet. Disposable paper liners are compostable and sometimes flushable, but if in doubt or worried about your septic tank, bin them rather than flush them down the toilet.

WHAT YOU CAN DO

- Choose reusable cloth nappies over disposables. Talk to friends who have used them and get their advice. Some people use disposables for the first few weeks of the baby's life when they are really tiny and then switch to cloth nappies when the baby is a little bigger; you're in more of a routine and you've got the hang of changing a nappy.
- Learn about the different types of cloth nappies, borrow some from a friend and find out what works best for you. Likewise, with liners, different options might work better for you at different stages of your baby's development or on days at home versus days out and about.
- Line-dry nappies rather than tumble drying; this will greatly reduce the carbon footprint of the nappies.
- Wash modern nappies at 60°C (no soaking needed) to kill germs for babies under three months' old and for those with skin problems; for older babies, wash nappy covers or lightly soiled nappies at 40°C. If in doubt check the advice given by the brand you use.
- Investigate a nappy laundry service if the laundry associated with reusable nappies is the thing that puts you off. Some of them rent you the nappies, take them away for cleaning and return them to you. They also swap up sizes as part of the package as your baby grows.
- Avoid nappy sacks. They are single-use, disposable and non-recyclable plastic. Just put the nappy straight in the bin without the extra plastic.

THE GARDEN AND GARAGE

A new petrol lawnmower used twenty-five times a year to cut the grass produces as much air pollution as eleven new cars being driven for one hour.

LAWNMOWER

Cutting the lawn can be a much-loved ritual or a tiring garden chore. Whatever your inclination, for many of us the sounds and smells of grass cutting are synonymous with summer and fine weather.

Before lawnmowers were invented, scythes were used to cut grass and only the most expert operators could cut lawns to an even height. That all changed in 1827 when Edwin Beard Budding, a Gloucestershire engineer, invented the pushalong cylinder lawnmower.

THE IMPACTS

A study commissioned by the US Environmental Protection Agency (EPA) in 2011 found petrol-fuelled lawn equipment – lawnmowers and strimmers – were responsible for 24–45 per cent of non-road petrol emissions in the United States. These harmful emissions include CO_2, which contributes to climate change, and volatile organic compounds (VOCs), which are thought to be harmful air pollutants.

A petrol lawnmower releases 40kg CO_2e and 24kg of VOCs and other air pollutants, such as nitrogen dioxide and carbon monoxide, per year (based on cutting the grass twenty-five times). The environmental footprint of a lawnmower comes in almost equal parts from making it and using it. So, using your lawnmower less can reap significant benefits for the environment.

It is not just the emissions that are a problem; petrol lawnmowers also contribute to water and soil pollution. The US EPA estimates that a staggering 64 million litres of petrol are spilled each year in the US simply refuelling lawnmowers and other gardening equipment – that's more than the oil spilled in Alaska by the Exxon Valdez oil tanker in 1989. The spilled petrol kills grass and anything else that is in the soil (earthworms, insects and microbes) and will take several weeks to recover as the petrol evaporates and is washed by rain into the soil.

Of course, there are other kinds of lawnmowers, too. Push lawnmowers, plug-in electric mowers and robotic mowers.

Push lawnmowers have the lowest carbon footprint of all as the energy consumed is merely the calories you burn while pushing! Well suited to smaller gardens, an electric lawnmower is easier to manoeuvre and has a lower carbon footprint than a petrol one, especially if it is powered by renewable energy.

And now robotic lawnmowers can mow the lawn while you do something else. They don't collect the grass but they convert the cuttings into a fine mulch that helps to fertilise the lawn. And as the grass is cut frequently, the clippings never amount to much, so you're not traipsing grass into your house. Robotic lawnmowers are electric so they have a lower carbon footprint than a petrol mower even if all the electricity on the grid is not yet renewable. Solar-powered robotic lawnmowers are not yet sold commercially – but undoubtedly they are the future of lawnmowing.

WHAT YOU CAN DO

- Get fit while you cut the lawn and use a push lawnmower! If you're buying a new one, choose a push lawnmower or an electric model over a petrol one. And be sure to find a new home for the old one or recycle it.
- Repair and look after the mower you own. Repair cafés (see page 187) and local small engine repair shops can help to maintain the engine and prolong the mower's life.
- Share or rent a strimmer, leaf blower or lawnmower with your neighbours. Less stuff means less environmental impact.
- Consider cutting the lawn less often to reduce the energy consumed by your mower.
- Leave wild areas in your lawn – there'll be less grass to cut and the local wildlife will thank you for it. Such wilderness areas help to promote biodiversity and provide food for important pollinators such as bees.
- Recycle old lawnmowers. Since these machines are made of multiple parts and materials, they can be broken down for recycling if you take them to the right place – you can't just leave them out with your household recycling. Push and petrol mowers can go in the scrap metal collection section of your recycling centre and the metal will be reclaimed and used again. Electric mowers go for WEEE recycling and so can be dropped off at your local recycling centre.

Children's author Roald Dahl wrote his much-loved books in a 1.8m x 2.1m garden shed he described as his 'little nest', with a woollen blanket over his lap to keep him warm.

GARDEN SHED

Whether it is a much-coveted man cave, a creative space outside the home for that new pottery hobby or a dusty place for storing garden equipment, the garden shed is a feature of many a garden.

Nowadays sheds can be workspaces, yoga retreats and even rooms for rent. An estimated two-thirds of people in the UK own a shed and 62 per cent say they would be deterred from buying a house if it didn't have a shed or a garden big enough to accommodate one.

Garden sheds are typically made of wood but can also be made from metal or plastic.

THE IMPACTS

The source of the timber in a wooden shed is important in terms of environmental footprint. According to the World Wild Fund for Nature (WWF), 18.7 million acres of forests are cut down annually, equivalent to twenty-seven football fields every minute. We have already lost half the world's tropical forest with impacts on wildlife, carbon storage and local livelihoods. This loss of forest is also contributing to climate change. Deforestation, agriculture and land use change account for just under one-quarter of global greenhouse gas emissions. Look for FSC-certified timber and you can have peace of mind that the wood your shed is made from comes from a sustainably managed forest.

To prolong their life, wooden sheds need to be painted regularly. Paints and wood-care products use solvents that evaporate into the air and help the paint or wood treatment to dry. Some solvents, called volatile organic compounds (VOCs), emit gases when you use them that can be harmful to health. Breathing in VOCs can cause irritation, headaches, difficulty breathing and damage the central nervous system and other organs. Old paint and varnish is toxic if not disposed of carefully (see also pages 168–169).

Plastic sheds are made from high-density polyethylene (HDPE) or polyvinyl chloride (PVC); they are very durable as well as being both water and UV resistant. Most sheds on the market are made from virgin plastic – requiring oil and energy for their production. But sheds can be made from 100 per cent recycled plastic.

All kinds of waste plastics (including low-density polyethylenes, HDPE, polypropylene from car bumpers, bins and bottles) are ground up, mixed and then fused together under high temperatures to transform it into a plastic 'wood'. This form of recycled plastic can be used for fences and garden furniture, as well as sheds, with the added benefit of being maintenance free.

At the end of your shed's life it is important to dispose of it carefully. The steel in metal sheds is valuable and can be recycled if you take it to a scrap metal dealer. Old wooden sheds that get thrown in a skip will be sent to landfill and contribute to climate change as they decompose. So, look for shed manufacturers that offer a 'take back' service for old wooden sheds when you buy a new one.

Good-quality plastic sheds should last somewhere in the region of twenty to forty years and afterwards could be ground down and recycled into another product. However, there are not currently facilities to collect and recycle such large pieces of plastic.

WHAT YOU CAN DO

- Look after the shed you have – it's the most environmentally sound approach. Repair it and maintain it so that it has a long life. For new sheds, research which type of shed suits your situation. Be realistic about your ability or willingness to maintain it. If you are unlikely to paint and look after a wooden shed, then consider steel or recycled plastic.
- Buy only an FSC-certified wooden shed (see also page 62). This certification means the forest the timber comes from is managed responsibly, protecting habitats, wildlife, local people and their livelihoods.
- Look out for sheds made from ocean plastic. The Ocean Recovery Project in the UK is collecting plastic litter from beaches and transforming it into marine waste plastic pellets that can be used to make garden furniture and, in time, fences and sheds, too.
- Recycle end-of-life wood and steel sheds and if looking to buy a plastic shed, ask questions about how it can be recycled, even if that is in tens of years' time. If enough people ask, a plan will have to be put in place by the producers and governments.
- Look for low-VOC paints to protect your shed and your health. Always dispose of old paint correctly (see also pages 168–169).

Additional fact

A survey carried out by Cuprinol in 2015 found that 5 per cent of shed owners worked from their sheds; this had increased to 13.8 per cent by 2017.

The average multipurpose compost, unless it's labelled 'peat free', contains between about 70 per cent and 100 per cent peat.

GARDEN AND POTTING COMPOST

If you are a gardener, you will almost certainly have a bag of garden or potting compost in your shed or stashed somewhere dry. But have you ever stopped to think where this amazing 'soil' comes from?

Historically, peat has been used in compost because it holds water and retains nutrients – just what gardeners need for superb results in their patch outdoors. Most garden compost is made from peat 'mined' from peat bogs, which are unfortunately a non-renewable resource.

The words 'peat bog' may not summon up the same romantic and natural visions as 'meadows', but bogs aren't empty expanses of waterlogged land, they are unique environments with amazing biodiversity. What's more, in the UK, peat bogs store more carbon than all of Europe's forests. The sphagnum moss in peat bogs accumulates layer upon layer over thousands of years to form moss peat soil. Sphagnum moss grows about 2–12cm per year to form peat up to six metres deep.

The top peat-exporting counties in the world are Canada, Germany, Latvia, the Netherlands and Ireland. Ireland is the source of most peat moss used in the UK. According to the Irish Peatland Conservation Council, up to fifty companies are currently mining peat moss from raised bogs in Ireland with minimal regulation and they estimate that over 60 hectares (120 football pitches) of new raised bog habitat was drained to produce moss peat products between 2014 and 2019, wiping out habitats for wildlife and contributing to climate change.

THE IMPACTS

Peat bogs are magnificent environments that nurture wildlife and preserve carbon. So, why are we digging them just up to sprinkle on our

A NOTE ON FERTILISERS

The overuse of synthetic fertilisers – in the garden and in the world at large – pollutes waterways, damages the soil and contributes to global warming. Using alternatives, such as homemade garden compost (see overleaf), the nutrient-rich liquid from a wormery, manure or seaweed, can help to condition the soil and add nutrients that benefit your plants.

gardens? Peat bogs serve several vital functions.

First of all, they are amazing stores of carbon. A 15cm-thick layer of peat contains more carbon per hectare than a tropical forest. Although peatlands cover only 3 per cent of the planet's land surface, they are what's known as carbon sinks – that is, undisturbed they store carbon forever, locking it away from the atmosphere. Peat bogs store at least one-third of the world's organic soil carbon. But as soon as the bogs are drained to extract the peat for compost or fuel, the stored carbon comes into contact with the air, triggering a decomposition process that releases carbon dioxide (CO_2) into the atmosphere. So, not only is a carbon store lost, but also greenhouse gases are released; it's a lose-lose situation. It is estimated that damaged peatlands release almost 6 per cent of global human-caused CO_2 emissions per year.

Second, peat bogs are home to incredible biodiversity, from plant-eating sundews and bog cotton to newts and butterflies. As they are drained for peat extraction, these special habitats are lost. Species such as the curlew, a threatened wading bird, are in sharp decline as peatland habitats are destroyed. There has been an estimated 97 per cent decline in their numbers in Ireland alone since the 1980s and at least 50 per cent decline in the breeding population in the UK over the past 25 years.

Third, bogs purify water and reduce flooding due to their capacity to absorb, hold and slowly release water. So, conserving bogs is a good defence against flooding, which is expected to become more intense with climate change, and it improves water quality. Preventing the drainage of remaining healthy bogs and rewetting damaged bogs is key to increasing our resilience to climate change.

WHAT YOU CAN DO

- Don't buy garden compost that contains peat. Look for compost that is marked as 'peat free' and examine the label closely. Don't be fooled by 'environmentally friendly' or 'organic' – neither of these ensure that compost is peat free. Peat-free compost is made from bark, coir (from coconuts), green waste, paper, leaf mould and sawdust, with added nutrients and water-retaining agents.
- Locate locally produced compost from industrial composters that process the contents of your food waste bin to produce peat-free compost.
- Make your own compost (see page 160).

DIY COMPOST IS WIN-WIN

Making good compost at home is an art and there are any number of online tutorials and courses delivered locally at garden centres and to community groups that can help you learn how.

Making compost saves you money:
There's no need to buy bags of garden or potting compost every spring when you already have a supply at home. When you reduce the amount of waste in your rubbish bin, you could cut your waste disposal costs. And with a plentiful supply of compost or the liquid produced from a wormery on tap, you won't need to use shop-bought synthetic fertiliser.

Making compost is good for the environment:
If you're composting food waste then less waste goes to landfill, so less climate-change-causing methane gas is released. By making your own compost, the peat moss of precious bog habitats is left intact, so that they can go on storing carbon, providing a home for wildlife, cleaning our water and protecting us from floods. What's more, composting at home means only organic substances are added to your veggie patch – no nasty chemicals in sight.

There are five essentials to making good compost:

1: The right mix of green and brown materials – Vegetable and garden waste on its own (green) will be too wet to decompose. It needs to be mixed with straw, leaves, wood shavings and shredded paper (brown) to get the right balance.

2: Moisture – All living things need moisture and that includes the organisms living in compost that decompose all the material. So, water your compost every so often to prevent it from drying out and cover it to reduce evaporation.

3: Air – Just as organisms need water to survive, they also need air; aeration is critical for decomposition. Having enough brown material and turning your compost helps the air flow.

4: Size of the heap – The size of a compost heap determines how hot it will get. Small heaps and compost bins don't get as hot as large, industrial compost heaps. As a result, home compost heaps just don't get hot enough to break down compostable containers.

5: Size of the pieces – Smaller pieces of waste break down more quickly. So, chop up bits of wood and hedge, shred paper and leafy material and make sure pruning offcuts are in small pieces as well.

Additional fact

In Ireland, 47 per cent of peatlands have already been destroyed by peat extraction and 92 per cent of raised bogs (the main source of peat for compost) in Ireland have been degraded. It's harder to restore a peat bog than replant a forest. Almost half of the endangered birds in Ireland rely on peatland habitats for their survival.

It is estimated that 27 million people in the UK, approximately 40 per cent of the population, do some gardening and growing. If they all buy just two plant pots a year – that adds up to 54 million plant pots.

SEED TRAY AND POT

A garden of varied plants, wild areas or a veggie patch is good for the gardener and the environment – storing carbon, producing food and attracting wildlife. Inevitably, though, gardening leads to a stash of plastic pots and seed trays.

Most plant pots are made of plastic and many are intended to be disposable, simply enabling you to get plants home from the garden centre.

THE IMPACTS

Plastic plant pots are made from high-density polyethylene (HDPE) or polypropylene (PP), both of which are recyclable. That said, not all kerbside recycling services will accept such pots due to concerns that they will not be completely clean. Also black plastics are notoriously hard to separate out at recycling plants, so these should not go in household recycling.

Recent consumer demand for alternatives to plastic seed trays and pots has led to a rise in popularity for compostable pots, but sadly many of these are made from peat (see also page 158).

WHAT YOU CAN DO

- Reuse the plant pots you have and pass spare ones on to other gardeners.
- Recycle clean, non-black plant pots if your waste provider collects them or find a garden centre with a 'take back' scheme.
- Look for plastic-free alternatives.
- Make your own seedling pots from toilet roll inserts or folded newspaper.

According to a 2017 study by Oxford Economics, 5.4 million acres of farmland across Britain are treated with glyphosate every year.

WEEDKILLER

As soon as the weeds start to poke out between the paving stones and along the side of the path in spring, fastidious gardeners, lawn enthusiasts and paving perfectionists reach for the weedkiller.

Claims such as 'Kills 99 per cent of weeds', 'Kills weeds and roots' share space on the weedkiller packaging alongside 'Safe for children and pets' and 'Will not harm your soil'.

Weedkillers work in different ways and have different uses – killing weeds in a lawn versus on a patio, for instance. Hormone-type weedkillers, made from ingredients such as ferrous sulphate, are selective and target broad-leaved weeds. So, when applied to a lawn they kill the dandelions without affecting the grass. Contact weedkillers or herbicides, on the other hand, are non-selective and scorch or burn off weed foliage. Systemic weedkillers differ from contact ones because the herbicide moves down though the plant to the roots, meaning that it is also effective at killing perennial plants.

Most of these are made from a chemical called glyphosate (see box, below), which is the subject of ongoing controversy.

KILLER CHEMICAL

First registered for use in the US in 1974, glyphosate is the most widely used weedkiller in the world. It accounts for about 25 per cent of all herbicides (another name for substances that kill plants) sold worldwide. It is licensed for use in countries including those in the EU, the US, Australia and New Zealand, with farmers claiming their yields could fall by up to 10 per cent without glyphosate.

THE IMPACTS

A weed is merely 'a plant growing in the wrong place' and in competition with crops. But weeds are important; bees and insects need wild plants like weeds for food.

This is a serious issue because insects pollinate about 75 per cent of the crops in the world as well as helping to fertilise soils and control pests. If their source of food disappears, they do too. Weedkillers are not the only threat they face; pesticides (which kill insects) are also used in gardening and agriculture and so also contribute to their decline.

Weeds can become resistant to herbicides over time, which means that new formulas have to be invented or more product has to be applied. There is a race on to see whether nature's ability to develop resistance can out-pace human innovation.

Weedkillers (and the companies that manufacture them) have been in the news in recent years due to the suspected long-term impacts on health. Pressure is mounting in countries around the world to ban the use of glyphosate.

At the time of writing, Sri Lanka, Columbia, El Salvador and Vietnam have bans in place, Belgium and the Netherlands have banned non-commercial use and France is committed to implementing a ban. Local authorities across Ireland and the UK have started to or are considering banning glyphosate.

In 2017 the European Union extended the licence for the use of glyphosate for a limited period of five years due to disagreement between member states regarding its safety.

WHAT YOU CAN DO

- Stop buying and using weedkiller. If you are a 'neat freak' and use weedkiller to keep paths and driveways free of weeds, try to think differently about the value of these plants for wildlife and humanity – remember that each is food for pollinators that enable the food we need to be grown.
- Apply mulch, woodchip or straw instead of weedkiller to keep down weeds. Pour salted boiling water on weeds in paving or sprinkle rock salt on paths in spring to keep weeds down. Pour vinegar on weeds to kill them – but do so carefully as it will kill anything green it comes in contact with.
- Do some weeding! Hands, hoes and forks can all be used to extract problematic weeds. Enlist your kids and get outdoors together.
- Don't cut your grass too short as this allows the grass to out-compete the weeds.
- Plant groundcover plants, such as periwinkles, that grow low over the soil and cover the ground to prevent weeds.
- Try having a 'no dig' garden. Instead of digging the soil to plant, cover the soil in compost, mulch, seaweed or straw and then plant. This technique reduces weeds, improves soil fertility, stores more carbon and increases yields.

Additional facts

Globally, 40 per cent of insect species are undergoing dramatic rates of decline.

Bees, ants and beetles are disappearing eight times faster than mammals, birds or reptiles.

The cordless drill is one of the most commonly purchased power tools in the UK with millions of them being sold each year, yet most people use a drill for only a few hours a year.

ELECTRIC DRILL

An electric drill makes hanging pictures, putting up shelves and screwing together self-assembly furniture a whole lot easier and faster. Electric drills were originally designed for industrial use, but when supervisors noticed employees taking drills home to use there, they recognised the potential for household use and do-it-yourself projects.

The first portable electrical drill was developed by Wilhelm and Carl Fein in 1895 in Stuttgart, Germany, and their company Fein continues to manufacture power tools to the present day. In 1961, Black + Decker introduced the first cordless electric drill, powered by a nickel-cadmium battery. And by 2005, Milwaukee, a US brand, had introduced lithium-ion batteries into cordless drills.

Nowadays, most drills have lithium-ion batteries which hold their charge more effectively than other batteries.

Modern drills come with an array of attachments – from sanders and wire cutters to brushes and grease guns. Try not to be tempted to buy too many specialist attachments as they often end up languishing unused in their box after their inaugural outing!

THE IMPACTS

Making a drill requires raw materials and energy. According to waste and sustainability charity WRAP in the UK, 91 per cent of a household drill's environmental impact occurs in the materials and processing phase with just 2 per cent occurring during its use. So, making drills last longer so that fewer of them are made and sold is a key way to reduce their environmental impact.

These days, it is not easy to repair a drill

if it breaks as it can be hard to find parts for household and lower-cost drills. You can buy spare parts for more expensive drills and this can extend their life. The battery in a drill can make up 40 per cent of the cost price of the drill, approximately £50, and when batteries stop working, consumers tend to replace the entire drill for a new model rather than just replacing the battery.

As most drills are now battery operated, the reuse of the battery is a major consideration in the environmental footprint of the drill. Lithium-ion battery recycling is still in its infancy, which is concerning because by some estimates there are only around 350 years' worth of lithium left in the world. Recycled lithium is currently up to five times more expensive than virgin material, which creates a disincentive for manufacturers to use recycled lithium.

Some companies, though, show that reuse of the batteries is possible. Aceleron repurposes vehicle, laptop and power-tool batteries for use in home-energy storage (such as those linked to solar panels on a roof). The company claims that approximately 70 per cent of the batteries it takes in can be repurposed in other applications.

Lithium-ion batteries are hazardous when the outer casing is damaged and the contents are exposed so they should never go to landfill. Battery recycling schemes in the UK, the EU, the US, Australia and New Zealand collect batteries for recycling to make sure that hazardous waste does not enter the environment and so that scarce and hazardous resources are reused.

WHAT YOU CAN DO

- Borrow a drill. If you only occasionally use a drill, borrow one from a friend or neighbour – just remember to return it. Use a sharing site to borrow or rent a drill.
- Choose tools with compatible battery packs to reduce the number of batteries needed.
- Take a broken drill to a repair café (see page 187) and see if it can be fixed. Look online for the spare parts you need.
- Sell parts from an old or unused drill online. Someone else can use them.
- Take old drills and batteries for WEEE recycling at your local amenity site or electrical retailer.
- Use a good screwdriver instead of an expensive electric drill for jobs such as assembling flat-pack furniture and for screwing in screws.

A REPAIRING PHILOSOPHY

The Restart Project in the UK is a social enterprise that organises events in local communities to teach people to repair their electronics and keep them longer, rather than having to throw them away. They estimate that the average drill is used for just 10 minutes in its lifetime. So, buying a good-quality drill that can be repaired, as well as sharing tools, such as drills, that are used infrequently, makes sense all round and reduces waste.

Additional fact

Initially the availability of electricity in the 1950s and then the rise in home ownership in the 1970s and 80s were key factors driving demand for electric drills. UK sales of electric drills rose from zero in the early 1950s to sales worth over £250 million by 2000.

It is estimated that plastic ropes and fishing nets make up 52 per cent of the plastic in what's known as the 'Great Pacific Garbage Patch' – a major ocean plastic accumulation zone in the subtropical waters between California and Hawaii.

ROPE AND STRING

Whether in a garden, up a mountain, in a sailing boat or at home wrapping a parcel, string and rope are endlessly useful.

Rope has been used for millennia to pull, drag, lift and hold things together. The earliest ropes were made from twisted grass, papyrus and reeds and used to haul rocks, lift logs and hold roof beams together. Rope and string were handmade until around 1850 when factories started to spin natural fibres, such as hemp, jute and sisal yarn, with machines. In the 1950s synthetic fibres were used to make a stronger, lighter rope that could be dyed different colours.

Synthetic ropes are important to climbers, for example, because they are lighter to carry than natural ropes, are stronger and have some elasticity. They are also popular for water-based activities as they don't hold water, some of them can float and they can withstand wear and tear. But they can be slippery and some types of synthetic rope degrade in sunlight.

Natural rope absorbs water, making it heavy, and will deteriorate more quickly than synthetic rope in damp and wet conditions.

THE IMPACTS

Natural rope and string are made from plants, such as hemp, sisal, flax, cotton and jute, all of which absorb CO2 as they grow. While all of these plants are renewable and can biodegrade, they do need land and water to grow and most industrial crops are grown using herbicides and pesticides, so that adds to their environmental footprint. Nevertheless, the overall impact of these natural fibres is far less than the synthetic alternative.

Rope and string made from synthetic materials, such as nylon and polypropylene, are ultimately

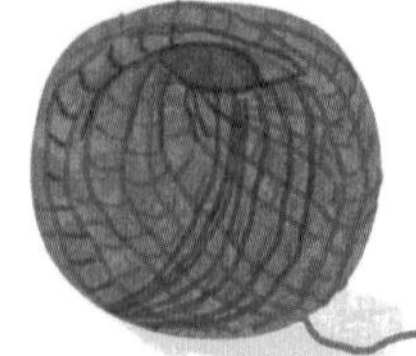

derived from oil, a non-renewable resource, and the manufacture of the plastic fibres requires a lot of energy and as a result produces carbon emissions. The qualities that make synthetic rope so popular, its strength and durability, are also the reason why old and waste rope is a problem. Synthetic rope is one of the most common types of litter found on coastlines and in the ocean, much of it from lost fishing gear. Not only do synthetic ropes not biodegrade, they are also difficult to recycle and while they are being used they shed plastic fibres into the water or the air (see also page 134). Old fishing nets made from ropes and fishing lines (also rope) are difficult to dispose of, so they typically end up in landfill or being incinerated.

Luckily solutions to the problem are being found with schemes to help fishermen recycle old nets and initiatives by divers to collect and bring to the surface ghost fishing nets for recycling and safe disposal.

Some companies are starting to collect and clean old ropes, including old fishing nets, to extract a recycled fibre that can be used in new ropes or to make everything from swimsuits (see also page 210) to skateboards.

Ropes can also be repurposed, so old climbing ropes, for example, can be turned into skipping ropes and dog leads.

CLEANING UP THE OCEAN

A company in Chile called Bureo collects disused fishing nets and ropes from coastal communities along the Chilean coast and then cleans and reprocesses them into plastic pellets. The pellets are then used to make sunglasses, skateboards and even a recycled version of the game Jenga.

WHAT YOU CAN DO

- Choose rope and string made from natural fibres. Natural string is good for cooking as it won't melt in the oven, discolour or add toxins to your food and it is compostable. Cotton is the best string for cooking and a roll of jute twine serves all purposes from gardening and parcel tying to children's craft projects. A gardener's string made from plant-based plastics (called PLA) is compostable in industrial composters but not at home. Rope made from natural fibres will not suit every need, but use it where you can.
- Look for brands of ropes for climbing, sailing and boating with a commitment to sustainability in terms of using recycled materials, planning for the end of life of the rope and using non-toxic dyes.
- Seek out mats, rugs, bowls, coasters and key rings made from leftover and waste rope, so nothing goes to waste. Look online for how to make your own rope mats and rugs.
- Put synthetic rope and string in the regular rubbish bin or find another use for them.

Additional fact

Approximately 640,000 metric tonnes of fishing ropes and nets is lost globally each year. Experts believe that fishing ropes and nets are the number one threat to seabirds, turtles and marine mammals who get tangled up and either drown or are injured.

It is estimated that 50 million litres of paint go to waste each year in the UK (enough to fill 20 Olympic-sized swimming pools). The average home is thought to have at least 17 half-empty pots stashed somewhere around the place, attracting cobwebs.

PAINT

Paint transforms buildings, protecting them from the elements and adding colour to indoor décor. Nowadays, almost every home renovation and do-it-yourself project requires paint.

After carefully choosing and testing to find exactly the right colour for your project, unfinished tins of paint tend to languish in sheds, under stairs and in garages, waiting, often in vain, to be used again to refresh a wall or repair a scuff.

People have been painting since 100,000BC using minerals, insects and clay bound with water, eggs and oil to add colour to drawings in caves and temples. And even modern paint has a simple recipe; it is made from three key ingredients:

1. Pigment to give it colour – these can be naturally derived from minerals, such as ochre, from plants or from coal tar, petroleum and petrochemicals.
2. Binder to glue or hold the pigment onto the surface being painted – these come from oil, which can be from a natural source, such as linseed, or from a synthetic resin, as well as containing acrylic polymers.
3. Solvent so that the liquid paint can dry quickly – most solvents, such as turpentine, white and methylated spirits, are made from petroleum.

Yet more additives such as biocides (to prevent the growth of bacteria and fungi), surfactants (to help the paint to spread) and driers (to help the paint dry), all add to the potentially toxic mix of ingredients in a can of paint.

That said, paint is an incredibly useful material, extending the life of buildings and furniture and is an important part of maintaining the things we already own and love.

THE IMPACTS

Some paint additives contain heavy metals – cadmium, lead, arsenic and zinc – which are hazardous to the environment and to health. While the levels of heavy metals used in paint are carefully controlled in the EU, the US and other developed countries, paints produced in developing countries have been found to contain dangerous levels of lead and cadmium.

Petroleum oil is an ingredient in many paints in the form of plastics – acrylic, polyvinyl alcohol (PVA) and vinyl. Plastics make the paint elastic, durable and mildew resistant, but at the same time they make it non-biodegradable.

Paint is classed as hazardous waste. Councils in the UK do not accept liquid paint as it is banned from landfill sites and cannot be recycled. In Ireland, approximately 30,000 tonnes of hazardous waste, including paint, from households and small businesses goes unreported and untreated every year.

Paints, varnishes and wood paints all contain solvents. Some of these volatile organic compounds (VOCs) emit gases when you use them that are thought to be harmful to health. In the air they interact with other particles to create ozone, which is a component of smog and can trigger asthma and damage lungs if exposed to them for a long time or in a poorly ventilated room.

WHAT YOU CAN DO

- Calculate how much paint you need accurately before you buy it. Apparently 75 per cent of people who do DIY simply guess how much paint they need and inevitably buy too much or not enough. Ask a professional painter for advice on how much you need or bring the dimensions of the room you are painting to the paint shop and they can advise.
- Store paint carefully to avoid it going off and it should last for up to ten years. Put the lid on properly to keep it from drying out and store where it will not be exposed to extreme heat or cold. For this reason, the garden shed is not the best place to store leftover paint.
- Try to use all the paint you have. If you don't need it yourself for touch-ups, pass it on to a friend or a neighbour or even a local community project (see page 170). The ends of tins of emulsion paint can be mixed together and used as undercoat.
- Look for the VOC globe symbol on paint. It tells you how much VOCs the paint contains. Buy natural paints that are VOC free, made from natural oils such as linseed, are plastic free (no vinyls and acrylics) and contain only natural pigments. There are lots of brands of good quality offering a wide range of colours.
- Minimise the amount of water or white spirit you use to clean your brushes and rollers by using a container to wash them in and old rags to dry them, which reduces the amount of waste you end up washing down the sink. If using the brushes again the next day, there's no need to clean them – just wrap them in an old plastic bag overnight so that they don't dry out.
- Dispose of paint correctly. First, you need to dry it out by buying a paint hardener or adding sawdust, sand or soil to the tin and leaving the lid off until it dries out. Then, ask your council where to take the paint for disposal. Solvent-based paint, paint thinner and white spirit are all classed as hazardous waste. Never put these in your bin.

GREENING UP PAINT

Given how toxic paint is, preventing it from becoming waste and making sure it is put to good use is a good idea. Projects are springing up around the world to do just that.

Community RePaint

Community RePaint in the UK helps to reuse paint by collecting leftover paint and redistributing it to people who need it and to community projects. Just go on to the website and register as someone in need of paint or with paint to donate. The network is made up of over sixty-five schemes and in 2018 redistributed over 317,600 litres of paint. Each scheme is unique but they all aim to brighten up people's spaces and their lives.

Paintback

Paintback in Australia is working towards a system for the responsible disposal and innovative reuse of paint. Established in 2016, it is an industry-led initiative to divert unwanted paint and packaging from ending up in landfill or the environment. Paintback is a charity that is funded by a 15 cents plus tax per litre levy on paint. Unwanted paint is dropped off at a Paintback location and transported to a treatment site where the packaging and waste liquid are separated. The containers are recycled and the waste paint is treated to recover the solvent and separate liquids from solids in order to minimise the waste going to landfill.

The average UK motorist will spend £168,880 on owning and running a car over the course of their lifetime.

Assessments of demand for electric vehicles (EVs) shows EV sales shifting from 2 million units in 2018 to 4 million in 2020, to 12 million in 2025, before rising to 21 million in 2030.

CAR

For many of us, getting around involves using a car. It is well established that petrol and diesel cars are bad for the environment, but substituting them all for electric cars isn't the answer either.

The first imaginings of a self-propelled vehicle dates back to the ancient Greek poem, the *Iliad*, and a self-moved tricycle.

There is no single origin story for automobiles as inventors around the world experimented with wind and steam to power carriages, with varying degrees of success. While steam dominated the scene from the late 1700s to the early 1900s, by the 1900s electric cars were on the rise.

At the beginning of the 20th century, 40 per cent of cars in the US were powered by steam, 38 per cent by electricity and only 22 per cent by petrol. Electric cars were silent, clean and fast compared with steam and the only thing holding them back was a lack of charging stations, as few buildings back then had electricity installed in them. The invention of the storage battery and the expansion of the electric network allowed the market for electric cars to grow and by 1912 electric cars were leading the way. Twenty companies were making them and in the US alone there were 33,842 registered electric cars.

So, what happened? Simple – the electric car didn't have the range of a petrol car and the invention of the storage battery, which had initially helped the expansion of electric cars, made petrol cars easier to start as there was no need for a wind-up crank shaft. So, by 1920, petrol cars had come to dominate the market and remain the most popular type of car driven today. But, thankfully, this looks set to change.

THE IMPACTS

All cars, whether they run on electricity, petrol or diesel, have an environmental footprint. Raw materials and energy go into the production of all vehicles.

In cars that run on petrol and diesel, the bulk (over 60 per cent) of their carbon footprint is generated when they are being used, reflecting the fact that fossil fuels are burned to power the car. In comparison, the making of the car equates to less than 20 per cent of the carbon footprint. The size of the car and the engine, the fuel used (petrol or diesel) and the age of the car all affect the emissions produced.

Petrol cars produce more CO_2 than diesel cars but diesel cars produce more air pollutants, such as nitrogen oxide and particulate matter. In 2018, tests carried out on sixty-one new diesel cars found that 80 per cent of them exceeded the legal and safe limits for nitrogen oxide. These emissions contribute to poor air quality and are one of the reasons why 2 million people in London are exposed to illegal levels of air pollution.

New clean air rules introduced in parts of London in 2019 (the so-called Ultra Low Emission Zone or ULEZ) have accelerated a decline in the sale of new diesel cars in favour of petrol and electric. By 2021, the ULEZ will cover the entire city.

There is a hot debate about how electric vehicles (EVs) compare with diesel and petrol cars in terms of their environmental credentials. Life cycle analysis of EVs reveals that the production phase (the making of the car and the battery) accounts for more of the carbon emissions than in a petrol or diesel car, but this is compensated for by lower emissions during driving. This analysis is based on an EV being charged by the current electricity mix (from nuclear, gas, oil, coal and renewables). Overall when conventional cars and EVs are compared, EVs have 20–27 per cent less impact. However, the footprint of EVs can be reduced further when charged with renewable electricity.

The lithium-ion batteries used to power EVs

VEHICLE-FREE SUNDAYS

Every Sunday, the congested city of Bogota in Columbia bans cars, trucks and vans from 75 miles (121km) of its roads and gives it over to cyclists, walkers and skaters – transforming the city centre into a peaceful paradise with a festival atmosphere.

have been the source of much investigation. Overall the battery is responsible for 14–23 per cent of the emissions associated with the disposal and recycling of an EV and 13–22 per cent of the total environmental footprint of an EV from cradle to grave.

A lithium-ion battery has a lifespan equivalent to almost 112,000 miles (180,246km) before it needs to be changed. The issue of what happens to used car batteries is crucial as they contain lithium and cobalt, both of which are expected to be in short supply by 2050. Much of the world's cobalt comes from politically unstable countries, such as the Democratic Republic of the Congo, and is associated with child labour. Lithium reserves are more plentiful than cobalt and are mostly found in Bolivia, Chile and Argentina. In this case, political tensions are a big factor contributing to concerns about availability in the coming decades. But there are also concerns about future deep-sea mining of cobalt and its impacts on the environment – so it is definitely an issue to watch.

The potential shortage of raw materials and concerns about pollution associated with redundant batteries is driving the focus on battery recycling. When a car battery is considered worn out in terms of powering a car, it can still hold up to 70–80 per cent of its charge and can also be used to store energy from wind and solar.

When the batteries are no longer able to hold a charge they can be recycled – about 58 per cent of used lithium-ion batteries were recycled in 2019. Recycling and reuse rates will continue to grow thanks to increasing demand for storage for renewable energy and the growing cost of the raw materials.

Biogas, made in anaerobic digesters from food and agricultural waste, is now being used to power buses and freight vehicles. This is a great way of solving a waste problem while creating an alternative fuel.

Considering most cars in the UK are parked 96 per cent of the time during a typical week, the cost of owning a car versus how much you actually use it does not make a lot of sense. In the UK in 2016 it cost, on average, £2,300 per year to run a small car and £8,900 a year to run a large car. And the costs of car ownership will remain high whether a car is petrol, diesel or electric. Electric cars cost a lot less to run than a conventional car but cost more to buy, so the cost of ownership remains high regardless – especially considering how little time an average car is actually on the road.

The aim in the future should not be to replace every petrol or diesel car on the road today with an electric-, biogas- or hydrogen-powered vehicle, it should be to have fewer cars, improved public transport and more widely available car-sharing facilities (see page 175).

CARS OF THE FUTURE?

So, are electric cars the future? Well, yes and no. The good news is that EVs are streets ahead of petrol and diesel in terms of air pollution, and when electricity is renewable with zero emissions, it will mean that charging and using an EV will be, too.

But there are other innovations to watch out for as well – hydrogen vehicles, for example, which hold great promise once the costs of making and fuelling them comes down. Biofuels such as ethanol made from sugars and grain will also play a role, although the biofuels of the future may be made from seaweed or agricultural waste.

WHAT YOU CAN DO

- Use car-sharing platforms, such as Go Car and Zipcar (see page 175), which give you access to a car but only when you need it. If you live in a city, using such a service is probably more cost effective than owning a car. The greenest thing you can do is to not own a car.
- Don't drive when you don't have to. Try to get into the habit of walking to the shop, to work, to school, to the pub or to your friends for dinner. It is a great way to fit some exercise and fresh air into your day. Did you know that just twenty minutes spent outdoors every day can make you feel happier?
- Share lifts to work, to school, to after-school activities, birthday parties and nights out. This means fewer cars on the road, so less congestion and fewer emissions. Liftshare is an online platform for businesses and individuals that helps match people to share car journeys.
- Maintain and look after the car you already have. A well-maintained car will produce fewer emissions, run more efficiently and last longer.
- Investigate buying an electric car when you're thinking of a new car. If most of your car trips are local then a new or, even better, a second-hand electric car could meet your needs. If you are getting a plug-in type of electric vehicle, apply for a government grant to install a home charger, make sure you choose an electricity supplier that provides 100 per cent renewable energy and switch your meter to a day and night meter to take advantage of cheaper rates when charging your car at night. You could even look into installing solar panels to charge your electric car, with your car battery effectively storing the solar energy your produce.
- Look into improvements in EV technology. If you travel longer distances and think electric cars are not an option for you due to their limited range, do some research as year on year the distance EVs can cover between charges extends and the incentives provided by governments to encourage their uptake become more favourable. Hybrid cars offer a good middle ground while improvements continue to made in terms of the range of EVs and a reliable and comprehensive charging infrastructure.

Additional facts

Cars are used for only 8 per cent of trips less than half a mile in distance but this rises to 76 per cent for trips between 2 and 3 miles (3.2–4.8km), a distance easily manageable by bike for many of us.

Driving on a heavily congested road can cause three times as many emissions as driving on a clear road – so traffic management and urban planning is an important part of the solution, too.

Car tyres and brake pads release particulates into the air during use, which contribute to air pollution – so the pollution from your car is more than simply what comes out of the exhaust.

ZIPCAR

Antje Danielson and Robin Chase met while their kids played in a park in Cambridge, Massachusetts in the late 1990s. As they chatted they came up with the idea for a mobility-sharing business that would reduce the need for cars owned by single individuals.

They started their business initiative – Zipcar – with one car in May 2000 and within four months had over 600 customers. The secret to the success for Zipcar was building a community of users that focused on the convenience and quality of the service provided. By prioritising these goals and messages to their users, rather than emphasising environmental or congestion-easing benefits, they made the car-sharing service a desirable and a fun experience.

Zipcar is now the leading car-sharing network in the world. By 2016 it had over 1 million members and operated in 500 cities and towns around the world.

Once you're a Zipcar member, you can book a car by the hour or the day. The whole process is managed through your membership card, online platform and an app. You just walk up to a car, tap your card, and drive away. It is as convenient as owning your own car because the cars are parked in neighbourhoods, at train stations, airports and universities.

The fee you pay is calculated per hour or day and covers fuel, insurance, parking and maintenance. The company estimates that it saves its customers up to £480–500 per month compared with owning and running your own car.

Worldwide, there are roughly 900 million dogs – in 2017 there were almost 90 million pet dogs in the US.

DOG POO BAG

Our four-legged friends benefit our wellbeing in many ways – they reduce stress, boost activity and improve heart health, to name but a few. It seems the only downside of having a dog is that you have to clean up after it. In the UK, dogs produce a total of 1,000 tonnes of excrement every day. That's a significant amount of poo to pick up.

Developed countries have laws about a dog owner's responsibility for picking up their dog's waste from public places. Public health campaigns emphasise the risk of pathogens in dog poo infecting people, with potentially serious results. Pet etiquette requires you to pick up and dispose of your dog's waste and, since you don't want any poo on your hands, that usually involves a 'poo bag' of some description.

THE IMPACTS

Dog poo is a biodegradable substance that will over time break down naturally. So, putting dog poo in a bag, while good from a public health perspective, is not good in terms of waste, as it encases a biodegradable waste product within a layer of plastic that will take hundreds of years to break down. In fact, bagged dog waste has potentially a greater environmental and aesthetic impact than uncollected dog waste, and plastic bags of poo hung on trees for collection and then forgotten pose a choking hazard to livestock and wildlife.

Many poo bags that claim to be planet-friendly or biodegradable don't actually break down within a year, and some take hundreds of years to break down. Compostable dog poo bags are available on the market – but since dog waste is not compostable in home or industrial composters (as it can't be used in compost that

PET ACCOUTREMENTS

Pets don't just create waste directly through their bodily functions, they also contribute to the waste stream through their food (see also page 179), toys and bedding. Pet toys and accessories, from cages to leads and coats to ball throwers, also add to the impact of owning a pet. Many are made from plastic and so lead to plastic pollution when they are no longer in use as most of them are made from non-recyclable plastics. Research has found detectable levels of lead and other toxins such as arsenic, chlorine and bromine in pet toys.

Here's what to get your dog instead of the plastic toys: deer antlers, felt balls, rope, or reused human items, such as an old leather shoe, or tied up pieces of old clothes, sheets and towels.

Look for second-hand pet equipment – everything from dog crates, cat boxes and kennels are available on buy-and-sell and swap sites online. Pets are expensive, so reuse old towels, blankets and cushions as bedding for pets.

will be used to grow food because of potential pathogens), this isn't really a solution.

A retired engineer, Ian Harper, has come up with a novel solution to the problem of plastic and dog poo. He has invented a dog-poo-powered biogas street lamp. Walkers in the Malvern Hills can use his free paper bags and then feed full bags into a biodigester. The methane produced in this mini anaerobic digester is stored and come dusk it powers a streetlamp. He reckons that ten bags of poo will power a streetlamp for two hours.

WHAT YOU CAN DO

- Get a dog-waste-only composter bin. It works like a miniature septic tank system, decomposing dog waste into a liquid that is absorbed into the ground. It's installed in a hole in your garden, and you add water and a special 'digester' liquid that contains enzymes and bacteria to help it break down.
- Dig a hole at least a metre deep in your garden away from any drains or water sources and bury your dog's poo there.
- Don't compost full dog poo bags, even if they are labelled compostable or biodegradable as dog poo is not safe to compost; they must go in the rubbish bin. Composting can't kill parasites and you don't want those in the compost you'll be putting on your garden.
- Clean up dog poo in your garden using a spade or a trowel to collect it and then put it directly in the bin. This reduces your use of plastic poo bags.

Additional facts

Over a quarter of the population of the UK has a dog – an estimated dog population of almost 9 million.

Australia has one of the highest rates of dog ownership in the world, with 40 per cent of households owning a dog.

Manufacturers of cat litter made from recycled paper claim it is three times more absorbent than clay, non-toxic and dust free.

CAT LITTER

Whether you live in a flat with no outside space or have only a small garden, the marvel that is cat litter can make the dream of having a pet real.

THE IMPACTS

Cat litter is usually made from clay or silica. Sodium bentonite clay is often used because it forms clumps when wet and is easy to scoop out for disposal. About 70 per cent of the world's bentonite comes from Wyoming in the US, where it is strip-mined. Such surface mining removes all topsoil to access the clay, destroys local habitats and scars the landscape.

An alternative cat litter is silica, which comes from quartz and is commonly found in sand. Urine and faeces stick well to silica but silica dust has been linked to respiratory problems in both cats and people. Used cat litter contains all kinds of bacteria and possibly parasites, so it needs to be disposed of carefully in the rubbish bin. It cannot be recycled and doesn't degrade, so it will just stay as it is in a landfill.

WHAT YOU CAN DO

- Choose a sustainable cat litter made from things that are often just thrown away, such as newspaper, woodchips and shavings, orange peels, corn cobs, wheat, bamboo and peanut shells. Many of these materials are waste products given another use and they avoid the need to mine natural resources.
- Never flush or compost cat litter as cat faeces may contain the eggs of *Toxoplasma* parasites. The eggs have tough shells that may not be killed in sewage treatment and could enter waterways and infect wildlife, such as otters. The composting process doesn't kill potential parasites and could expose people and other animals to toxoplasmosis, which can be dangerous for pregnant women and immune-suppressed individuals.

Pet food is responsible for around a quarter of the environmental impact of meat production when you take into account land use, water, fossil fuels, phosphates and pesticides.

The Dutch company Protix 'grows' 1 metric tonne of insects to make pet and animal feed in six days using food waste and a land area of only 20 square metres.

PET FOOD

In the past dogs and cats were fed on offal, offcuts and scraps – but the trend now is for bespoke food to cater for a rising prevalence of allergies and problems.

Owners are increasingly insisting on human-grade meat for their pets, meaning that pets eat about 20 per cent of the world's meat and fish.

THE IMPACTS

As more people around the world have pets, demand for pet food contributes to the need to produce more meat and catch or raise more fish. More livestock inevitably means more greenhouse gas emissions. Demand for fish for pet food contributes to overfishing, and demand for low-cost fish can lead to poor working conditions for staff on fishing boats.

To compound the problem, pet owners are overfeeding their pets, leading to obesity, health problems and big vet bills. The worldwide prevalence of pet obesity lies between 22 and 44 per cent, and rates seem to be rising. In addition to ill health, overfeeding your dog can reduce its life expectancy by up to two years.

WHAT YOU CAN DO

- Ask your vet how much food your dog or cat needs and stick to the guidance to avoid feeding them too much.
- Research sustainable pet food. A company called Yora is making pet food from 40 per cent insect protein mixed with oats and potato. The insects are the larvae of black soldier flies (*Hermetia illucens*) that are reared in the Netherlands from waste food and vegetables.
- Dogs and cats are not vegetarian by nature but they do need a balanced diet. Get advice from your vet before you consider a plant-based diet.

THE WORKPLACE AND SCHOOL

7 million disposable coffee cups are used every day in the UK, or 2.5 billion every year.

22,000 takeaway coffee cups are thrown away every hour in Ireland.

TAKEAWAY CUP

Coffee 'on the go' is a modern craze. Today's grandparents didn't grow up drinking coffee from plastic cups – in fact, many of them find the whole experience of drinking from a takeaway cup pretty unpleasant compared with a nice crockery mug or a genteel cup and saucer.

The origins of the disposable coffee cup are as recent as the 1960s, with the invention of the polystyrene cup. So, this is a truly recent phenomenon.

THE IMPACTS

The problem with takeaway coffee cups is that we use them for such a short time and then throw them away. 'Paper' cups are misleading as they are lined with plastic, usually polyethylene, to make them waterproof. This means they are not recyclable and so have to go in a rubbish bin bound for landfill or incineration.

The plastic lids are recyclable but only if they are clean – so any residue of milk or coffee has to be washed off first. You may have read statistics such as only 1 in 400 coffee cups in the UK is recycled – but as they stand, most takeaway coffee cups are not recyclable in standard recycling facilities.

If a coffee cup becomes litter the paper part will decompose but the plastic liner will merely break down into microplastics, which risk being ingested by wildlife and people if they end up in the food we eat and the water we drink. The realisation that these plastic-lined cups are not biodegradable and that they pose a risk to the environment sparked a move towards reusable cups and compostable cups from 2015 or so.

Keep Cups were the original reusable cup for takeaway coffee and they were first sold in Melbourne, Australia in 2009. If you have a reusable coffee cup and you use it once a day, seven days a week, you avoid the use of 365 cups per year and if you are a two coffees a day person that is 730 disposable cups avoided annually.

But it is not just the waste that's an issue, it is also the raw materials needed to produce the cups. Takeaway coffee cups are made from paper, which comes from trees, and plastic, which is made from petroleum with added chemicals to make it soft and pliable. It also takes energy to make the cups, which produces carbon emissions and the emissions increase further when you take into account the trees cut down to make the paper. Using a reusable cup has a considerably lower carbon footprint than either a standard paper cup or a compostable cup. In fact, using a reusable cup rather than a disposable one for a year saves the equivalent carbon to that absorbed by five trees in a year.

Compostable coffee cups are becoming more common as an alternative to plastic-lined paper cups. They are not the full solution, however, as they are designed to be single use and takeaway. The absence of appropriate bins on the street, however, means compostable cups rarely make it into a brown bin for composting, and in a bin they are just regular rubbish.

WHAT YOU CAN DO

- Create a new takeaway coffee ritual. Buy a reusable cup you love and use it every day. The choice between glass, bamboo, plastic or stainless steel is less important than how regularly you use it. So, choose a cup you like to drink out of, remember to carry it with you, and keep in mind that every time you use it you are reducing your carbon footprint and the waste you produce. Opt for a thermally insulated cup if you like your coffee super-hot. If you are always on the move, look for coffee cups will 'no spill' lids or a collapsible silicone cup.
- Sit down and enjoy your coffee from a normal or reusable cup.
- Keep any compostable cups (after drinking the coffee) and take them home or to the office to go in a brown (organic) bin (for food waste collection). Don't put it into a regular rubbish bin or it defeats the purpose of it being compostable. And don't put them in the recycling – they can't be recycled.
- Clean and then recycle the plastic lids of regular takeaway coffee cups.

Additional fact

If everyone in Australia switched from disposable cups to reusable cups it would save carbon emissions equivalent to flying a Boeing 747 for 100,000 hours.

Over 4 billion of the 7.5 billion (53 per cent) people living on Earth today use the internet, and in 2017 it was estimated that 1.3 billion people worldwide owned a personal computer.

COMPUTER

Having access to a computer and the internet makes it possible to connect with people, set up a business, learn, study for a degree, access information to protect your rights and make informed decisions.

According to the Organisation for Economic Co-operation and Development (OECD), 91.7 per cent of UK homes had a functioning computer in 2017. Similar levels were seen in other countries around the same time – 83.8 per cent for Ireland, 82.4 per cent in Australia and 78 per cent New Zealand. However, in 2015 research by the American thinktank the Pew Research Center revealed stark inequalities in computer ownership between developed and developing countries. At that time 80 per cent of citizens in the US owned a personal computer, while in Indonesia it was 13 per cent and in Uganda it was only 3 per cent.

The rise of the personal computer has transformed lives since the first forays into computing in the 1940s to the arrival of desktop computers in the 1970s and 1980s. Innovation in the computer sector has been staggeringly fast and with that has come pressure to buy the latest model and have the best technology. As a result, consumption of computers is big business even with recent minor sales declines as tablets and smartphones replace some of this market. Over 58 million computers (desktop and laptops) were made in the first quarter of 2019 alone. That's a lot of computers especially when you consider that the average lifespan of each computer is only three to five years.

THE IMPACTS

Most of the environmental impact associated with computers is due to manufacturing, and although computers are becoming smaller in size, they still require more than ten times their weight in chemicals and materials during production. Metals such as copper, lead and gold are used in a variety of computer parts – lead might be used for solder or radiation shielding, gold for pin plating and copper as a conductor. Aluminium, magnesium, silicon, zinc, cobalt, nickel and iron are also commonly used to make hard discs.

Heavy metals, such as mercury, are used in circuit boards and switches, along with cadmium in batteries and chips. Making computer parts also requires very rare materials: ruthenium, which is rarer than gold and platinum and is used in high-performance hard discs; and hafnium, a material used in processors that may run out completely in ten years at its current rate of consumption.

Another raw material, coltan is a black tar-like mineral that is used as a heat-resistant powder to hold a high electrical charge. It is used in capacitors in electronics ranging from laptops to phones. Eighty per cent of the world's coltan is in the Democratic Republic of the Congo where men and boys extract the mineral from the Earth in dangerous conditions, for little pay. Many of the mines are controlled by militia and the sales fund war rather than benefitting the people of Congo.

Making a computer also requires energy as does powering up and using it. The carbon footprint of the laptop I am typing on, from cradle to grave, is 286 +/–50 kgCO2e – equivalent to driving 701 miles (1,128km) in a car. Of that total, 76.2 per cent comes from the manufacturing stage, 20 per cent from use, 3.6 per cent from transportation and 0.2 per cent from end-of-life recovery (recycling).

Seeing as the manufacturing of computers is so energy intensive, any effort companies can make to use less energy and reduce their carbon footprint makes a difference.

Once a computer is brought home from the shop, it uses, on average, 746 kilowatts (kW) of power each year, requiring more power than a fridge (500kW/year). Leaving it switched on or on standby when not in use increases your carbon footprint and your energy bills (see also page 89). In general, a desktop computer uses more energy than a laptop computer, which in turn uses more energy than a tablet.

With models of computer only a few years old being classed as obsolete, the issue of electronic waste or e-waste is an ever-growing issue as consumers feel the pull to buy the latest model. E-waste needs to be handled carefully to minimise the harmful materials within the computer and to reclaim the valuable components and materials for reuse.

According to the United Nations, 44.7 million metric tonnes of e-waste were generated worldwide by 2016, and only 20 per cent was recycled through appropriate channels. This equates to 6.1kg of e-waste per capita annually in 2016. Europe is the second-largest generator of e-waste per capita with an average of 16.6kg/inhabitant; that said, Europe has the highest collection rate at 35 per cent due to the fact that they have e-waste legislation and incentives in place.

The Americas generate 11.6kg/inhabitant but collect only 17 per cent of the e-waste for safe disposal and recycling. There is still controversy about exporting e-waste to developing countries where the facilities and regulations for safe disposal and recycling are not in place.

Recycling your computer is a good thing to do in terms of both preventing pollution and maximising resources. According to the Environmental Protection Agency in the US, recycling 1 million laptops could save enough energy to power 3,657 US households for a whole year. A study in Mexico in 2014 found that the use of recycled materials versus virgin materials had tenfold benefits in terms of cutting emissions and threefold benefits in terms of reducing human exposure to toxins.

Heavy metals pose a health risk if they escape into the environment if a computer is not disposed of carefully at the end of its life. Increasingly plastics are used in computers and the main impact of plastic is when the computer becomes waste.

WHAT YOU CAN DO

- Set your computer to automatic sleep mode after 15 minutes of inactivity so that if you are away from your desk it powers down to save energy.
- Turn off your computer, monitor and printer when you're not using them. To make this more convenient use a power strip to turn everything off at once.
- Computer screens use most of the energy, so turn them off if you are taking a break from your computer and turn down the brightness to a level that works for you.
- Repair your computer. Don't buy a new one without checking to see if it can be repaired.
- Recycle electronic equipment – never put it in the rubbish. Take your old computer to the electronics section of a recycling centre or to a retailer that offers a 'take back' recycling service. Some companies will even cover the costs of you mailing an old computer back to them for recycling.
- Look at your reuse options. Is there a local organisation that could reuse your computer?
- Ask questions about where your laptop is recycled. Check out the StEP Initiative online to see where computers are recycled. Before you sell or recycle your computer you need to remove everything from the hard drive. Online tutorials explain how to do this or ask at your local computer shop.
- Do some research before buying a new computer. Take a critical look at the brand you are interested in before you buy (see also page 251). See how transparent they are about how they manufacture the computer, their CO_2 targets, their use of renewable energy and their commitments related to taking back old computers, managing hazardous waste and recycling and reuse. Check, too, how different models compare in their energy efficiency and carbon footprint.
- Switch your electricity provider to a renewable energy provider to reduce the carbon footprint of running your computer.

Additional facts

According to the Environmental Protection Agency, computer monitors can contain up to 3.6kg (8lb) of lead, which is toxic to humans.

Laptops are kinder to the planet than desktops, partly because they contain less of the harmful substances and partly because they use less power.

A study in Nigeria showed that, in 2015/2016, EU member states were responsible for about 77 per cent of used electronic equipment imported into the country for recycling.

REPAIR CAFÉS

Worldwide, manufacturers have embraced what's known as 'planned obsolescence' – deliberately designing products to fail prematurely or become out-of-date, so that they can sell you a new model. The average life expectancy of a new washing machine is eleven years but if it has a fault after eight years, the recommendation will be to replace it.

A study conducted in ten repair cafés in the Netherlands assessed 715 repairs to identify the most common items. The most commonly repaired objects were bicycles, vacuum cleaners and kettles, with an average success rate of 70 per cent.

In response to this 'throw away and buy a new one' mentality, and the growing electronic waste problem, a movement of repair cafés has sprung up around the world – in countries including the UK, Ireland, New Zealand, Australia and the US.

Started in 2009 in the Netherlands by Martine Postma, repair cafés are free meeting places, where volunteer specialists can help you to repair a broken household object. The cafés have all the tools and materials needed to carry out repairs if you know how or you can get the help of an expert. They are all voluntary and can help you to fix anything from clothes and furniture, to electrical appliances, bicycles, toys and crockery.

In addition to extending the life of household goods, repair cafés help people to pass on lost knowledge about how to repair things, plus they reduce waste and save carbon emissions, as the energy and raw materials needed to produce a new product are saved. They are also great places for people to meet and create hubs for sustainability in communities.

At a repair café in Dublin over half the objects brought in for repair are mobile phones, tablets and computers.

The US EPA estimates that 1.6 billion pens are thrown away in the US every year. Unfortunately old pens and lids are commonly found on beach cleans where the action of the sea and the sun breaks them down into microplastics that can enter the food chain.

PENS

We have come a long way from relying on a quill and ink to write with, but at what cost? In the last 70 years, pens have become disposable items and are now largely made of plastic. How many pens are hidden in your drawers?

BIC pens were among the first disposable ballpoint pens on the market in the 1950s. Disposable pens have a plastic case and top, a plastic tube to hold the ink, a textured tungsten carbide ball and the point that holds the ball is made of brass. All of which adds to the mix of materials, making such pens hard to recycle.

THE IMPACTS

The biggest issue is the fact that pens have become so cheap and disposable that they are overused and frequently end up as litter. BIC sold its 100 billionth disposable ballpoint in 2005 – that is a lot of pens and a lot of plastic.

As with other plastic items, disposable pens are produced from oil along with additives such as phthalates (see page 26). The solvent in ink may be harmful to health and can contribute to air pollution when exposed to sunlight.

WHAT YOU CAN DO

- Treat yourself to a good-quality, refillable pen. You can now buy refillable ballpoint pens, ink pens, whiteboard pens and highlighters. Buy the refill ink at the same time.
- Use a pencil instead. Look for pencils made from recycled newspapers and paper, or even recycled jeans. Make sure wooden pencils are made from FSC-certified sustainable wood.
- Choose an 'inkless' metal pen that never runs out; it has a metal tip that lays down a grey line that looks like pencil but doesn't smudge or erase.
- Turn down free pens.

CENTRALISING CREATIVE ENDEAVOURS

Lots of businesses have leftover packaging, end-of-line items, offcuts and excess stock that get thrown away because they are not needed or there is no space to store them.

As rubbish, these items take up space in landfills or get incinerated when they could be having another life as any number of things – such as part of a set on a stage, as a sculpture or a craft made as a Mother's Day gift. That's where the not-for-profit organisation ReCreate comes in.

ReCreate takes in unwanted items from business – such as promotional pens and pencils, the colourful plastic inserts of spools of thread used for weaving or end cuts of aluminium foil used to package coffee – and makes them available to schools, families and community groups for creative projects.

Their Dublin warehouse is packed with all types of creative gems from plant pots to buttons and bottles to rolls of material, as well as pens, paint, glue and string. This projects encourages creativity through reuse.

Anyone can become a member to access the materials, attend workshops and get advice. ReCreate also organises workshops for groups and schools from disadvantaged communities who get to access the art supplies free of charge.

Their aim is to inspire curiosity, creativity and care for the environment, all while diverting waste from landfill and incineration.

Worldwide an estimated 1 million ink cartridges are thrown away every day.

In the UK, 55 million toner and ink cartridges are disposed of annually.

PRINTER AND INK

The ability to print from a home or office computer has revolutionised how things get done – from printing your own boarding pass or cinema tickets to photographs and bank statements.

The first computer printer was made in 1938 and a dry printing process later known as Xerox came online in 1971, setting the stage for laser printing. It took a couple of decades more, though, for printers to become small and affordable enough for home use.

The 2018 internet access survey in the UK revealed that about 40 per cent of adults own a desktop computer. Not everyone who owns a computer will have a printer, but even if half do, that is a lot of printers – about 29 million in the UK alone – and each printer needs electricity and ink in order to work.

Printers come in two main types for home and office use: inkjet and laser. Inkjet printers use the ink contained in cartridges and a series of tiny guns (jets) to fire dots of ink at the paper, creating an image or words from many small dots. A laser printer uses an electronic circuit to process the data from a computer or tablet and then a laser beam scans over and back across the page creating a pattern of static electricity that attracts and holds the toner (held in a replaceable cartridge) on the page.

THE IMPACTS

The environmental impact of a printer comes from the manufacturing, transport, use and disposal of the printer and its associated ink cartridges. Research published in 2012 on the life cycle of an inkjet printer found that the paper use stage of printing is the most environmentally impactful, followed by manufacturing and electricity consumption. And, for inkjet cartridges, the end-of-life phase, which covers recovery, recycling and disposal,is a significant source of environmental impacts.

Inkjet printers use ink cartridges filled with liquid and dyes, while laser printers use toner cartridges that contain a fine powder, mostly of finely ground polyester. Most inks are designed to be non-toxic and the main ingredients are water, ethylene glycol and alcohol. However, colour pigments, resins, black carbons and traces of heavy metals (such as lead, cadmium, mercury and chromium VI) can cause skin or eye irritation and can leak from computers and ink cartridges, causing pollution if they are not disposed of carefully.

The energy involved in manufacturing just one brand-new toner cartridge for a laser printer is estimated to result in 4.8kg CO_2 (equivalent to driving an average car for about 160 hours), while a recycled or remanufactured cartridge emits an estimated 2.4kg CO_2. Overall a remanufactured cartridge (prepared for reuse and refilled) is best, saving energy and resources and costing the consumer less. Approximately 20 per cent of all cartridges sold around the world are remanufactured.

Of course, the use of the printer over its lifetime also consumes energy; if this electricity is produced from fossil fuel, and most still is, then there's the emission of greenhouse gases that cause climate change to contend with, too. Small quantities of ozone, volatile organic compounds (VOCs), which speed up drying, and tiny particles are released during laser printing. If concentrated, for example in a small room with poor ventilation, these particles could have impacts on human health. To date, these particles have been classed as a nuisance rather than a hazard by the United States Environmental Protection Agency.

A staggering 1 million laser and inkjet cartridges are thrown away each day and many of them end up in landfill, where they will take hundreds of years to break down, releasing chemicals into the environment in the process. It is estimated that only about 30 per cent of toner cartridges are currently recycled around the world. This is such a waste, as 97 per cent of the materials used in both inkjet and laser printer cartridges can be recycled, reducing the need for virgin raw materials and energy needed to manufacture new cartridges.

Obsolete printers at the end of their lives contribute to the growing e-waste problem worldwide (see also pages 72 and 75), when if collected they can be recycled or even remanufactured to be used again.

Until recently, old printers were often sent to developing countries for recycling, but without regulation and enforcement the printers can end up being dumped or incinerated. Tighter regulations in the European Union and countries including Australia and New Zealand now provide for special recycling services for e-waste including printers, and many manufacturers now provide a 'take back' service for empty toner and ink cartridges.

WHAT YOU CAN DO

- Print only when you need to (at home but also at the office) – this saves energy, paper, ink, money and carbon emissions.
- Check your print settings – set your default to print both sides of the page and/or 2 pages per sheet, make sure you don't have colour as your default print setting, opt for draft or greyscale to reduce ink use and use the 'print preview' function to avoid mistakes.
- Refill ink cartridges. Although the printer manufacturers don't encourage us to do this, it reduces waste and saves money.
- Make sure you recycle all your used toner and ink cartridges so they don't end up in landfill. Some manufacturers, retail shops and charities offer a 'take back' service for used cartridges. Just check the manufacturers' website or ask in the shop where you buy them.
- Turn off your printer when you are not using it. Left on standby it is consuming energy and adding to your electricity bill and your carbon footprint (see also pages 60 and 89).
- Choose an EU energy-certified A+ or more rated inkjet printer over a laser jet as they are more energy and cost efficient. Do some research before you buy and look for manufacturers that produce energy-efficient printers, offer a 'take back' service for cartridges and printers at the end of their lives, make new printers and cartridges from recycled materials and produce ink that is free from hazardous substances.
- Look for inkjet printers with a refillable ink reservoir, called an 'eco tank' by some manufacturers. This cuts out the need for cartridges altogether and you just fill the tank as needed. Another advantage is that you are less likely to have a cartridge run out just as you are about to print something important as it is easy to see the ink levels in the tank. Manufacturers claim that you can use 80 per cent less ink in a refillable tank printer rather than one that uses cartridges.
- Investigate the idea of managed print services at work – which streamline and centralise printing in an office with energy-efficient printers and enforced rules on printing like double-sided and black-and-white printing.
- Take your broken or old printer back to the retailer or to a WEEE collection point for recycling. Some manufacturers do offer their own 'take back' services – so do check their website.

THE MOVE TO RECYCLING PLASTICS

Hewlett Packard have recently announced that more than 80 per cent of their ink cartridges and 100 per cent of HP LaserJet toner cartridges are now manufactured with recycled content. Plus, they have used over 227,000kg of ocean-bound plastic to make Original HP ink cartridges.

Dell is using reclaimed ocean plastics in the packaging for its computers. Moulded trays are now made from 25 per cent ocean plastic and 75 per cent recycled plastic, so no virgin materials are used and, what's more, the trays are fully recyclable.

It is estimated that the average US office worker uses up to 10,000 sheets of printer paper per year and it takes up to 13 litres of water to make one A4 sheet of paper.

COMPUTER PAPER

'Think before you print' and 'Please consider the environment before printing'. We may see these common reminders at the bottom of emails, but do we ever take heed of them? Often not. The promise of a paperless office seemed tantalisingly close a few years ago but, in fact, now people print off emails and information from websites so, in fact, the volume of paper printing has gone up, not down.

Paper use in workplaces from hospitals to universities varies according to the type of business. Legal companies, for example, are often required to have printed copies of documents, while software companies are likely to be almost paperless.

One way or another, people have been using paper for a long, long time. Stone tablets gave way to papyrus, which was first used as a base for writing in Mesopotamia in 3000BC. In China, people wrote columns of text on thin strips of bamboo from 1500BC, which likely led to the invention of the first paper. According to Chinese history, a eunuch in the Han court named Cai Lun invented paper in 105AD. Nowadays, China, the United States and Japan are responsible for the production of half of the planet's paper and North America uses more paper than any other region in the world.

THE IMPACTS

Although paper is made from a renewable resource – trees – it still has a significant environmental footprint. The World Wild Fund for Nature (WWF) estimates that the paper industry uses 40 per cent of all the timber that is traded internationally. Trees come from forests, which cover 31 per cent of the land on Earth, purify the air and the water, store carbon and provide livelihoods to 1.6 billion people.

Approximately 2.6 billion metric tonnes of carbon dioxide (CO_2) is absorbed by forests every year. That's almost one-third of the CO_2 emissions released from burning fossil fuel to create energy – so forests are vital in protecting us from the worst impacts of the pollution that human activity creates.

Yet we are cutting down our forests at the rate of twenty-seven football pitches per minute, which actually causes emissions as each tree cut down is a store of carbon that is lost and this contributes to climate change. Overall 25 per cent of global greenhouse gas emissions come from cutting down trees, clearing new land for agriculture and for building and expanding towns and cities. About half of these emissions are due to deforestation and the decline in the health of our forests.

The paper industry is also the largest user of water per metric tonne of finished product.

HOW PAPER IS MADE

Paper is made by a process called pulping, where fibres, usually from wood, are mixed with water (a lot of water, approximately 189 litres of water per tree, as well as pulp-promoting chemicals) to make what is called a suspension. The liquid is then drained off to leave a mat of fibres or pulp. If the paper is being bleached, it happens at this stage before the pulp is then pressed and dried to make paper. Nowadays we have lots of choice between virgin and recycled paper, FSC-certified paper from sustainably managed forests, recycled paper, bleached and unbleached paper – all compatible with laser and inkjet printers.

Between 300,000 litres and 2.6 million litres of water are used to make 1 metric tonne of paper. The amount varies according to the processes used. As well as using vast volumes of water, making paper also involves chemicals – from the pesticides used in forests to the 200 or so chemicals used in the pulping process to help extract the fibres from the wood. All this combines to make the paper industry the fourth-largest polluter of surface water globally. What's more, chlorine is used to whiten paper which, if it escapes into rivers and streams, can cause harm to animals and plants living in the water.

Manufacturing paper also requires a lot of energy and produces greenhouse gas emissions. Paper- and pulp-making in the UK is responsible for 6 per cent of industrial greenhouse gas emissions, primarily due to the use of fuel for machinery and electricity. There is great potential to reduce emissions in paper-making by increasing efficiency in the manufacturing processes and by switching to renewable energy. The European Union predicts that the European pulp and paper industry can reduce its emissions by 63 per cent by 2050 compared with 2015, through efficiencies and the use of clean energy.

Recycling paper has multiple benefits. First, it saves trees – somewhere between seventeen and twenty-four trees per metric tonne of paper – and it turns a waste product (used paper) into new paper, which avoids unnecessary landfill and incineration. It also saves energy – recycling paper uses somewhere between 27 and 70 per cent less energy (depending on the type of paper product) than making virgin paper. It also used 30–40 per cent less water than making virgin paper. This all helps to reduce air pollution by 73 per cent and water pollution by 35 per cent and it creates jobs in collecting, separating and reprocessing the waste paper into new paper.

Yet globally it is estimated that just 37 per cent of paper is recycled, which is a wasted opportunity when you consider the benefits from reducing waste to protecting forests. Every piece of paper can be recycled five to eight times before the quality starts to deteriorate.

WHAT YOU CAN DO

- Reduce your paper use – only print if you need to (see also pages 190–191).
- Buy unbleached recycled paper labelled '100 per cent post-consumer waste' to tell you that is made from paper that was already used, rather than from offcuts or from manufacturing waste. The paper will be white but not as brilliantly white as bleached paper. It uses a process that de-inks the used paper rather than bleaching it.
- Reuse paper that is printed on one side for taking notes, for writing shopping lists or for using in kids' projects.
- Recycle paper – remembering to tear out any plastic windows from envelopes; you can recycle recycled paper, too.
- Have separate recycling bins for paper in your office, so that it will not be contaminated by other waste.
- Remember shredded paper can also be recycled and if you use a shredding company – at home or in your workplace – then make sure they recycle your paper.
- Look for FSC-certified chlorine-free paper. It can be labelled as ECF (elemental chlorine free, so no chlorine is used to bleach the pulp) or, even better, TCF (totally chlorine free, which means no chlorine is used in the pulping or paper-making).

WHAT'S IN AN ENVELOPE?

Most envelopes are made of paper and as such are recyclable. Padded paper envelopes sometimes contain a layer of plastic bubble wrap, which means they are no longer recyclable. The plastic envelopes that online shopping is sometimes delivered in are often not recyclable.

The most confusing envelopes by far, though, are the most popular ones for official correspondence – those with the transparent plastic windows. The windows can't be recycled, so you'll need to tear these off and put in the bin before you recycle the rest.

Printing directly on envelopes rather than on labels reduces waste, although labels can be helpful to enable you to reuse padded envelopes, for example.

Choose envelopes made from recycled paper and without plastic windows. You can get padded envelopes with paper padding that are recyclable. Some envelopes have pull-off 'peel and seal' strips to close them, which create unnecessary waste as the strips are not recyclable. Choose gummed envelopes instead.

Additional facts

The pulp, paper and printing sector was responsible for 5.6 per cent of global industrial energy use in 2014.

Nearly two billion hectares of degraded land across the world could be restored and reforested to absorb carbon and provide a solution to climate change.

In 2019 a Scottish school was the first in the UK to ban glitter and one-quarter of nurseries in the UK have said they would like a ban on glitter for environmental reasons.

GLITTER, GLUE, STICKERS AND CRAYONS

When I was a child, arts and crafts involved glue, crayons, paint and various things gathered from around the house, such as cereal boxes and toilet roll inserts. Nowadays, the fashion is to buy art supplies – glitter, stickers, plastic googly eyes, foam shapes and lots more.

Let's take glitter, for example. As far back as 40,000 to 10,000BC, people used mica flakes, akin to fool's gold, to reflect light and add a twinkle to cave paintings.

Henry Ruschmann invented plastic glitter in 1934. Since then we have found endless uses for glitter, from greeting cards and eyeshadow to wrapping paper and nail varnish. We even put glitter in pens, crayons and glue.

THE IMPACTS

Most glitter is made from multilayered sheets of plastic, colouring and a reflective material, such as aluminium, titanium dioxide or iron oxide. It is a microplastic and washes down the drain after 'make and do' projects and, because of its size, it easily passes through wastewater treatment plants and enters the environment where it poses a threat to wildlife.

Glitter is so light it can also be transported in the air and in recycling plants it causes problems when it causes machinery to clog, and it contaminates wood pulp since it is so small it is hard to wash out before making new paper.

To get glitter to stick to a homemade card or drawing you need glue. Glue is made using solvents that are air pollutants – so, glue with lower levels of solvents or that are solvent free are better both for health and for the environment.

Such glues and adhesives are also used to make stickers, which are often embellished with glitter or have a plastic coating. Stickers are not recyclable, and neither is the paper they come on or the sticker books that sometimes accompany them.

If you have kids, chances are you have drawers or boxes filled with crayons. Crayons are great for small hands learning how to draw but, unfortunately, they are not great for the planet. Crayons are made from paraffin wax (a by-product of petroleum) that it is mixed with dyes and other additives (such as glitter)! Paraffin wax is not biodegradable and the different chemicals added can cause pollution. When you think of all the crayons kids use at home and in school and add to that the crayons given out for kids to play with in restaurants and in lucky bags – that is a significant amount of crayons.

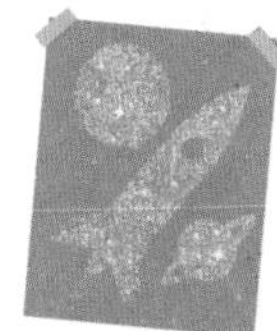

Some crayons now come inside a plastic tube and they twist out to be used, a little like a propelling pencil. While these are nice to use, they add yet another elements of waste to a crayon. The hard plastic tube is not recyclable since there is likely to be crayon residue left inside along with the metal spring mechanism that makes them twist.

GIVING CRAYONS A NEW LEASE OF LIFE

The Crayon Initiative in the US collects unwanted crayons and the ends of crayons from restaurants, schools and families and melts them down to make new crayons. This stops the crayons from becoming waste and creates new crayons that are donated to arts projects in children's hospitals.

WHAT YOU CAN DO

- Say 'no' to all glitter. Try salt, rice or lentils on cards and art projects instead.
- Buy plastic-free glitter that is biodegradable. Be careful because they are not all 100 per cent plastic free and if they have some plastic they are not biodegradable. Only buy biodegradable glitters that have been independently checked and certified – it should say this on the label or on the manufacturer's website. Mica is a natural alternative to glitter but mining it is linked to human rights abuses, so ethical businesses are no longer using it.
- Look for solvent-free glue. Empty glue bottles and glue stick covers are not usually recyclable as they will contain residue of glue. TerraCycle has a specialised glue packaging recycling scheme, so go online and look for a drop-off point near you.
- Look out for paraffin-wax-free crayons made from plant oils and waxes.
- Avoid crayons in plastic tubes – it is just more unnecessary plastic and waste.
- Look for a crayon recycling programme you can donate to.
- Avoid stickers, plastic eyes, foam shapes and beads where you can.

Additional facts

More than 12 million crayons are made in the US every single day. And 20,000–34,000kg of broken crayons are discarded in landfills throughout the country annually.

In 2018 sixty-one music festivals across the UK banned glitter.

'How to make slime' was the most searched 'how to' phrase in 2017 – leading Google to call 2017 the 'Year of Slime'.

SLIME

Kids love slime. It is fun to make and the squelching and squishing calms kids of all ages down. There are thousands of 'how to make slime' videos online; many of them have been watched hundreds of thousands of times.

One specialist slime YouTuber – yes, there is such a thing! – has over 657,000 subscribers. A search for slime on YouTube in 2018 would have revealed 29 million results. Slime even made up one-quarter of Argos's Arts & Crafts range in 2018.

Slime is made from a reaction between PVA glue (a polymer) and a slime 'activator', such as borax (the base ingredient of many laundry detergents) or contact lens solution, to make a stretchy substance. Other ingredients are added to give different properties – shaving cream can make it fluffy while body lotion makes it smooth and soft. Once the texture is sorted, then comes the colour and added extras – glitter, plastic balls and beads, paint, foam sponges, eyeshadow, nail varnish... I could go on.

THE IMPACTS

For most people the joy of slime is in the making, getting the texture right and the gloopy and popping sounds it makes as you mix it. That means that lots of slime is made, played with for a day or less and then thrown away.

The problem with slime is that it is essentially plastic, non-biodegradable and not recyclable. Washing old slime down the sink means it will break into smaller and smaller pieces to form micro- and nanoplastics.

PVA glue is made from polyvinyl acetate, a flexible, water-soluble substance that is non-toxic unless you eat it. Borax – the activator in slime – is also known as sodium borate, which is designed to be a cleaning product first and foremost, and not a regular plaything. Borax can irritate skin. Since young children put almost anything in their mouths, close supervision is needed as borax is not meant to be eaten. Repeated exposure to borax, by handling it or breathing it in, is also not recommended. So, letting our kids make regular slime is not the wisest thing – no matter how much they love it.

The extra ingredients added to slime – glitter and plastic balls and beads – compound the problem of throwing away yet more plastic (see also pages 196–197). All of the additives to the slime pose their own individual risks to wildlife and as a combination they are potentially lethal.

WHAT YOU CAN DO

- Say 'no' to any more plastic-based slime and tell your children why it is not a good idea for them or the environment.
- Make PVA- and borax-free slime from cornflour, food colouring and water (see box below for recipe and instructions). Or try a mix of shampoo or washing-up liquid and cornflour – you can still use your hands and experience that squishy and stretchy feel.
- Try edible slimes (see recipe below) – your kids will love these. Go easy on the edible slime as it is full of sugar!

MAKE YOUR OWN SLIME

Schools and play therapists recognise the therapeutic value of squeezing and squishing to make slime. So try these plastic-free and edible versions.

PLASTIC-FREE SLIME

What you need:
Cornflour
Water
Food colouring (if you wish)
A large bowl and spoon

How to make it
Tip out some cornflour into a bowl. If you are using food colouring, add a small amount of it to a bowl of water rather than directly into the mixture; don't add too much or you will stain your hands. Mix approximately 2 parts slime with 1 part water, stirring all the time. This takes some experimentation to adjust the quantities to get the texture you want. As you stir, it will come together in a lump and then you can add more water or cornflour to get the consistency you want. Then, stretch, squish and enjoy.

Store in a container in the fridge; it will last a few days. Add a little water to reinvigorate it.

EDIBLE SLIME

What you need:
Jellies or marshmallows
Food colouring (if you wish)
Icing sugar
Sprinkles

How to make it
Melt the marshmallows or jellies in the microwave for about 30 seconds or in a bowl over a pan of simmering water. Stir in a few drops of food colouring, if using it, then add some icing sugar a spoonful at a time, stirring all the time until you get the consistency you want. Only handle it when it is cool.

For texture and visual feast, you can add some sugar sprinkles or other edible cake decorations and work them into the slime.

Warning: Keep all slime away from furniture – whatever it is made from!

LEISURE TIME

It takes 23kg of petrochemicals and minerals plus 925 litres of water to make one typical yoga mat.

YOGA MAT

Ask busy and successful people how they manage their hectic lifestyle and chances are yoga will often emerge as a key strategy. Practising yoga can relax us, slow us down and tune us into the natural world – a positive approach to activity, health and the planet.

Now a modern popular activity, yoga has been around for millennia and has its origins in India. Stone-carved figures of people in yoga poses have been found in the Indus Valley dating back to 3000BC. Yoga blends physical exercise with spiritual and meditative practice and has many different disciplines and adaptations. Modern yoga incorporates more physical poses than in the past when it was often conducted sitting or standing, making a mat increasingly important to the practice.

THE IMPACTS

Attributing environmental impacts to a spiritual practice so deeply connected to nature seems contradictory, but most people today practise yoga on a mat made of polyvinylchloride, or PVC – essentially plastic.

PVC mats are cheap, durable and sticky, so your feet don't slide when holding a pose. They are also cushioned to provide protection and comfort. They roll up easily when not in use and can be wiped clean when needed. A PVC mat typically costs around £20 while mats from other materials (such as cork or latex) could cost £100 and more.

Although PVC mats, therefore, have lots of positives, PVC is a synthetic resin, so essentially it's a plastic. PVC is transformed from a tough, rigid material (as used in pipes) to a softer and more flexible form through the use of additives such as phthalates (see also page 26). PVC can

UPCYCLED YOGA MAT ANYONE?

Suga, a company in the US, collects old wetsuits via a network of surf shops and turns them into yoga mats – giving the wetsuits a new life. One of their products is essentially a yoga mat for life and if it gets damaged or needs to be replaced, they will do so free of charge and take back your old one. Similarly, an Irish company called The Upcycle Movement collects old wetsuits and turns them into yoga mat straps and carriers.

be unstable when exposed to light and heat. Stabilisers made from metals, such as lead, barium, calcium or cadmium, are added to the PVC to prevent this from happening, but add potential pollutants to the materials in your yoga mat.

Because PVC contains the additives needed to make it stable and soft, it is not easy to recycle. So, your average yoga mat is ultimately destined for landfill or incineration, if it's not reused or repurposed in some way. Yoga mats are not biodegradable, and they release dioxins into the atmosphere when incinerated and in landfills they produce a toxic leachate that has to be carefully managed.

WHAT YOU CAN DO

- Take care of your yoga mat and keep it as long as you can. PVC mats are durable, so if you look after it well it should last decades.
- Look for alternatives to PVC when buying a new one. Natural rubber or latex mats are available and have good grip (although not suitable for people with allergies to latex). Check, too, that the rubber comes from a sustainably managed forest. Cotton, cork and jute mats are also available. In fact, there is a lot of choice now, so you can weigh up sustainability and comfort with durability and grip and come to a decision you are happy with, in terms of how it's affecting the planet.
- Pass on old yoga mats. The more your mat is reused and the longer it stays out of landfill or incineration the better.
- Look for yoga mats made from waste, such as old wetsuits collected from surf shops. Old wetsuits also make great yoga carry straps.

Additional facts

Two billion people around the world practise yoga – and if even half of them do so with a mat, that is 1 billion yoga mats.

Yoga's popularity is evidenced by a 2017 study by Lancaster University that found that yoga is one of the top fifteen most popular words in Britain, alongside words such as 'Facebook' and 'twitter'.

Old yoga mats can be given a new life – as a roll mat for camping, an insulating wrap for cold drinks or a portable seat for concerts and festivals.

In 2017, 63 per cent of men and 58 per cent of women took part in sports, with running being the sport of choice for 15 per cent of people.

Worldwide, 25 billion pairs of running shoes are sold per year (34 million per day), making shoes the most significant contributor to the resource consumption and environmental impact of running gear.

RUNNING GEAR

Like almost every other sport, running now requires 'gear': running shoes, running socks, running shorts, moisture-wicking tops, reflectors for running in the dark, a watch to time your run and a smartphone to play music and track your run. All that gear is going to have an impact.

Most runners are not consuming the latest gear for the sake of it; they do need to look after their feet and bodies with trainers to support their gait and to avoid injury and they need clothes that are flexible and prevent chaffing. Fortunately, the range of sustainably produced kit is ever growing.

Modern sportswear fabrics are synthetic and are designed to wick away moisture, reduce friction, control odours, give maximum flexibility and provide a good fit. On the whole, natural fibres have fallen out of favour as they tend to get heavy when damp or wet and can cause friction and rashes. But we may have written them off as sports textiles too soon – new innovations mean natural fibres can perform in tough conditions.

THE IMPACTS

In 2008, the magazine *Runners World* calculated the annual carbon footprint of a typical competitive runner (not an elite or professional runner). They calculated everything from the carbon footprint of socks, shorts and T-shirts, the energy required to wash and dry gear as well as the travel to races and workouts, to arrive at a total carbon footprint per runner of 2,472kg CO_2e (equivalent to driving 6,044 miles/ 9,727km) in an average car).

Of the different elements the most significant contributors were travel to events, running shoes and washing and drying gear. In 2012 the *Washington Post* went a step further to calculate the carbon footprint of a runner doing a marathon in terms of the CO_2 emissions associated with the food they ate and the CO_2 they expired. They found that running more slowly produces fewer CO_2 emissions than running quickly!

The manufacture of a typical pair of running shoes is responsible for 14kg CO_2e, equivalent to driving 34 miles (55km) in an average car. Running shoes or trainers are made of many different materials and are not recyclable. Most trainers are made from synthetic materials and plastic, so they consume oil and don't biodegrade. Keen runners can get through three or more pairs of running shoes in a year, so finding a longer-lasting shoe can reduce your environmental footprint considerably.

Running clothes are also synthetic so when they are washed they release plastic microfibres into the wastewater system and, eventually, into rivers and the sea. Anti-odour clothing often contains silver nanoparticles that act to kill bacteria and smells. Some studies have reported that nanosilver can leach out of textiles when they are washed. It is thought that nanosilvers can accumulate in the environment and pass up the food chain, posing risks to health; while current levels are unlikely to cause chronic damage, they are being monitored by scientists to prevent any longer-term harms, especially as the clothes are disposed of at the end of their life.

Most running gear is not recyclable and ends up at a landfill site or in an incinerator, so get the most wear from your gear before you relegate it to the bin.

PLOGGING

Plogging started in Sweden and combines running or jogging with picking up litter – it encourages runners to play their part in cleaning up the environment. You can plog on your own or find a plogging club or event near you by searching online. The plogging craze is growing right across the world – there are over 83,000 plogging posts on Instagram alone!

WHAT YOU CAN DO

- Pass on gear you no longer use or that doesn't fit. The most sustainable gear is the gear you already own.
- Choose eco-conscious running gear. There are clever people looking for sustainable solutions to the synthetics that have dominated running apparel in recent decades. In particular look for items that are long lasting, made from sustainable fabrics (for example, bamboo and organic cotton mixes), use recycled materials in the fabric and that offer a 'take back' service for old gear. The brand Sundried, for instance, makes high-quality, quick-drying running gear out of coffee grinds and recycled plastic.
- Source sustainable socks. The Teko brand, based in Italy, makes running, hiking, cycling and skiing socks from merino wool, recycled polyester (from plastic drinks bottles) and synthetic fibres (from recycled fishing nets). They also use non-toxic dyes as well as minimal packaging.
- Look for running shoes made from marine plastic or recycled plastics – there's now plenty of choice.
- Don't overwash your gear – washing and drying adds to the carbon footprint of your clothes. Try airing things outside and washing less often. Put synthetic items in a Guppy bag (see also page 109) to prevent microfibres escaping from your machine along with the wastewater.

Over half the people on Earth know how to ride a bike, and it is estimated that there will be 5 billion bicycles on the planet by 2050.

BIKE

Getting about under your own steam on two wheels is one of the greenest forms of transport, in terms of its impact on the environment (only walking beats it). Plus, you get fit without even trying. Result!

Cycling for leisure and as a form of transport is on the rise, as the sport increases in popularity and as cities evolve to encourage more people to switch from cars (and public transport) to bikes. Demand for bicycles is growing at a rate of more than 100 per cent per year.

Luckily the modern bike has evolved from the cumbersome wooden contraptions of the early 19th century and the first real bicycle – the Penny Farthing, with its huge front wheel and smaller back one. The bike as we know it today came about in 1885 when John Kemp Starley, an Englishman, invented what he called a safety bicycle with two wheels the same size and a chain to drive them.

Nowadays, Lycra-clad groups zooming through the countryside on club rides or commuting to work in cities is a common sight. And as cyclists embrace all that nature has to offer, they can be key advocates for sustainability.

THE IMPACTS

If you're a cyclist, you are already doing good for the planet and for your health. Unsurprisingly, cycling 2 miles (3.2km) produces much less CO_2 than driving; for instance, driving produces 0.88 kg CO_2 while riding a bike clocks up a mere 0.017 kg CO_2. That said, the equipment you use has an impact.

Bicycles are made from multiple materials – aluminium, steel, carbon fibre, natural rubber, and synthetic rubber primarily, with smaller amounts of silicon, iron, copper, manganese, magnesium, chromium, zinc, titanium and nylon. Aluminium and steel both need to be extracted from ores (bauxite and iron ore) mined from the ground and processed with lots of energy to transform them into bicycles.

A life cycle analysis of bicycles made in Bangladesh in 2019 revealed that bicycles made from aluminium have the highest environmental footprint, followed by steel, while carbon-fibre bikes have the least. Unless, that is, you opt for a bamboo bike – an innovator in Ghana is making lightweight, sustainable bikes from bamboo!

Most of the carbon footprint of a bike is attributed to the extraction and preparation of the materials and the manufacture (67 per cent), while assembly is responsible for only 5 per cent and use and maintenance is responsible for 15 per cent – this includes changing the tyres and doing repairs.

Synthetic rubber is made from oil and chemicals while natural rubber comes from trees, both of which impact the environment, albeit differently. Bike tyres are usually made from synthetic rubber and steel and can be recycled if sent for specialist tyre recycling. Using recycled rubber in tyres can reduce the use of petroleum and bring down the carbon footprint of tyres. Inner tubes are also made from synthetic rubber, though they can be repaired if punctured.

Most bike helmets are made from a polyvinylchloride (PVC) or polycarbonate plastic shell with the all-important shock-absorbing-foam layer underneath. This foam is usually an expanded polystyrene, made from small plastic pellets that are inflated, moulded and then steam fused into the shape required. Polystyrene is not biodegradable and is only recyclable in specialist facilities.

If you have a bike accident and hit your head, the polystyrene in your helmet absorbs the shock but cannot bounce back – the upshot is you need a new helmet and the old helmet has to go in the bin as it is can't be recycled. However, safety first!

A SUSTAINABLE BIKE HELMET

The EcoHelmet combines the need for safety and convenience with sustainability. It is designed for spontaneous users of bike schemes, who often don't wear a helmet. It is made from waterproof recycled paper assembled in a honeycomb design to make a helmet shape.

Still in the testing stage, its award-winning design claims to absorb shocks as effectively as polystyrene. It folds flat when not in use and is 100 per cent recyclable.

WHAT YOU CAN DO

- Look after your bike: repair it, maintain it and fix punctures. A well-looked-after bike can have a long life.
- Give away an old bike. Look out for projects such as Rediscover Cycling in Ireland that collects and revamps old bicycles for a new owner.
- Take old tyres to be recycled with car tyres.
- Look for new tyres made from natural rubber and recycled synthetic rubber.
- Research before buying a new bike or helmet. Ask about its materials and look out for minimal or recyclable packaging, which demonstrate a brand's commitment to sustainability.
- Look out for brands of cycling gear that are following the trend set by others of making apparel from recycled and marine plastic (see also pages 204–205 and 210).

Every 10,000 metric tonnes of recycled nylon saves 70,000 barrels of crude oil and avoids 57,000 metric tonnes of CO2e emissions (equivalent to the energy needed to power 6,826 homes for a whole year). Recycling nylon reduces its global warming impact by up to 80 per cent compared with making it from scratch.

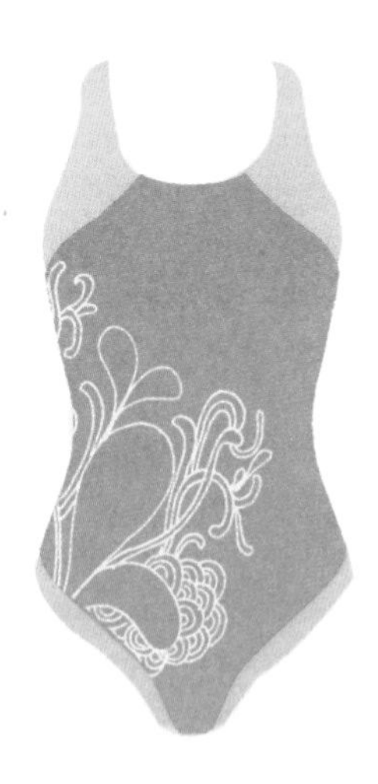

SWIMSUIT

Ever wondered what those sleek suits Olympic swimmers wear to make them more hydrodynamic are actually made of? One thing is for sure, they are a lot easier to swim in than the swimsuits of yesteryear, which were much more about modesty than performance.

The first bathing suits were designed in the second half of the 19th century and their main purpose, for women in particular, was to conceal the body. Typically a dress-type garment over bloomers, they were designed not to rise up in the water (some even had weights in the hems) and were made of material that would not become see-through when wet. They were heavy and cumbersome as a result. In the early 1900s, swimwear for men and women became more athletic, revealing knees and arms and allowing swimming as a sport to develop.

Nowadays, swimsuits are made from synthetic (or plastic) fabrics: nylon, polyester, polybutylene terephthalate (PBT) and Lycra (also called elastane). Polyester is more resistant to chlorine and sun damage than nylon and while Lycra gives a good fit, it does deteriorate after exposure to chlorinated pools. PBT is a texturised polyester and it holds up well for sea and pool swimming and is both stretchy and close fitting. An estimated 65 million tonnes of these plastic-based garments are manufactured every year.

If you're a swimmer, you probably also have some goggles and perhaps a swimming hat. Goggles are made from rubber and silicone as well as the polycarbonate hard plastic used to make the lenses. Modern swimming hats are made from silicone or synthetic fabric.

THE IMPACTS

All swimsuits designed for swimming, as opposed to sunbathing, are made from synthetic fabrics. This means they are derived from oil, a non-renewable resource, and just like the plastics of swimming hats and goggles, they will not biodegrade.

The added challenge with swimsuits is that as they are worn and washed they shed microfibres into the washing machine and that wastewater feeds into natural waterways and contributes to ocean plastic pollution.

Swimsuits made from recycled nylon from fishing nets and water bottles (see also page 210) help to reuse otherwise waste materials but it doesn't solve the problem that synthetic fabrics shed microfibres in water. To find a solution to that problem we have to return to natural fibres.

The company Outerknown, co-founded by surfer Kelly Slater, set out to find a way to manufacture a new generation of swimwear – one based on natural materials that wouldn't contribute to marine pollution. They have made the world's first 100 per cent merino wool swimming trunks that are all natural and biodegradable, and there aren't any plastic microfibres to shed into the water.

Called Woolaroos, the board-short-style trunks are breathable and quick drying and don't hold odours. By using a traditional material and stretching and spinning the yarn with new technology, they have made it water-resistant and machine-washable, without the need for chemical additives.

Currently, swimming hats and goggles are not generally recyclable and neither are swimsuits. So, although many swimmers love splashing about in the sea – sadly what they wear could be contributing to ocean pollution.

WHAT YOU CAN DO

- Choose a swimsuit made from regenerated nylon made from ghost fishing nets or from recycled plastic (see page 210).
- Sign up for a TerraCycle Eyewear Zero waste box at your swimming club or pool. These boxes can be used to collect old goggles (and reading glasses and ski goggles) for specialist recycling by waste organisation TerraCycle.
- Wash your swimsuit in a Guppy bag (see also page 109) to capture microfibres and reduce the risk of them contaminating wastewater.
- Rinse swimsuits, goggles and hats in fresh water after swimming to make them last longer.
- Pass on old or no longer worn swimsuits to a friend or a charity shop until a 'take back' system is created.
- Upcycle – you can use old swimsuits as stuffing for cushions or to make headbands and stress balls.
- Want a plastic-free swimming hat? Go retro and try one made from natural rubber or latex. The manufacturers say the thicker rubber keeps your head warmer, especially good for those swimming in the sea.

SWIMWEAR FROM THE SEA

OceanPositive and Bakoto are among the growing number of swimwear manufacturers using recycled nylon, or Econyl, to make a full range of swimwear.

Econyl is a fabric made from ghost fishing nets (nets lost at sea that continue trapping and killing marine life as they are carried on the currents or lie on the sea bed), plastic bottles and even carpet. The ghost nets are collected from the sea by divers and NGOs such as The Ghost Fishing Project and Ghost Nets Australia.

In 2018 over 100,000 metric tonnes of fishing net were collected from the ocean – approximately 10 per cent of the total lost and discarded – and turned into recycled nylon. Econyl is fully recyclable and companies that use it are working on ways to take back old swimsuits and give them another life.

OceanPositive and other brands such as Julienne that make sustainable swimwear also pay attention to their packaging. When you order online your swimsuit is posted to you in a non-plastic bag made from cassava starch and other renewable resources. The bag is compostable and is even safe to be eaten by animals, should it escape into the environment.

Swimsuit companies are also spreading their message to the industries they collaborate with. For example, OceanPositive set up Mission 2020, an initiative to engage with organisations working in the diving industry to change their business practices to preserve the oceans. Their first activity is a pledge to eliminate the use of single-use plastics in members' businesses by 2020.

The company Athleta, now part of the GAP group, makes 85 per cent of its swimwear from recycled materials. By using recycled nylon, they estimate they have diverted 72,264kg of waste from landfill, equivalent to the weight of 2.4 humpback whales.

There are over 7,000 gyms in the UK, and one in seven people is a member of a gym. Every one of those gyms has fitness machines, such as treadmills, stationary bikes, stair climbers, cross trainers and rowing machines, that are powered by electricity as well as the gym's TVs, lights and air conditioning.

GYM EQUIPMENT

Whether you go to the gym a few times a week or jump on the treadmill or exercise bike at home to get some exercise, gym equipment is increasingly part of people's everyday lives.

The whole notion of gyms, fitness and body image originated in ancient Greece, where gymnasia such as the Academy and the Lyceum in Athens were places where men met to socialise, train and gain social status.

The first commercial gyms were opened in Paris and Brussels in the 1840s by a Frenchman called Hippolyta Triat. One of the first gyms in the UK was opened by German strongman and showman Eugen Sandow in London in 1897 called the Institute of Physical Culture. Sandow's fascination with the human body, fitness and a strong physique spread with his gyms.

With modern life becoming increasingly sedentary, the focus on activity for physical health has become well recognised and promoted by health professionals around the developed world. So, it's good to see that by the 20th century going to the gym was something everyone could do.

THE IMPACTS

A typical gym is full of electrically powered machines: treadmills, rowing machines, static bikes, elliptical trainers, steppers, cross trainers, the list goes on. Add to that the TVs, fans, air-conditioning units, sound systems and array of lights and it's easy to see that the energy consumption of a gym is significant.

Depending on the size of a treadmill's motor, the weight of the runner and the speed it is set to, it uses between 300 and 1100 watts of energy per hour. The costs of this depends on the electricity provider, but based on an average set-up in 2018 in the UK it cost 18 pence per hour to use. So, if you run for three hours a week that equates to 54 pence per week, or £28 per year. Now, scale that up with the numbers of machines at a typical gym, the hours they're open (some are even open 24 hours a day) and the gyms round the country, and you can see that the energy use soon mounts up!

It is estimated that gyms and sports facilities in the UK spend £700 million per year on energy, emitting 10 million metric tonnes of CO_2, equivalent to 2.6 coal-fired power plants operating for one year.

As with many modern machines, gym equipment continues to use energy while on standby (see also page 89) and the energy used to cool the air or heat the water for a hot shower afterwards all contributes to the carbon footprint of a workout.

A 2014 study looking at air quality in gyms in Portugal discovered high levels of air pollutants, called volatile organic compounds (VOCs, see also page 168–169). The VOCs typically come from the building's materials, floor coverings, gym equipment, cleaning supplies and hand sanitisers – features of gyms the world over. Poor ventilation adds to the problem, especially at busy times when people in classes kick up and inhale lots of dust. The researchers expressed concerns about the concentrations of dust and chemicals in the air and the risks posed to people with asthma and respiratory illness.

GREENING POWER FOR GYMS

Eco-friendly gym equipment (harnessing the energy produced during workouts to drive the machines or to power lights and TVs at the gym) is on the rise. Many green gyms – outdoor gyms in parks – use such technology (see opposite).

The average person creates up to 300 watts of energy during a workout, so channelling this energy back into the equipment saves money and reduces carbon emissions.

WHAT YOU CAN DO

- Exercise outside when possible and reduce the carbon footprint of your workout to zero. Even 20 minutes a day in the outdoors improves your mood.
- Look for an energy-efficient or a self-powered piece of gym equipment that uses the energy you produce while exercising to run the machine. Most weights-based gym machines do not use energy and new manual versions of equipment, such as treadmills and rowing machines, are available.
- Research into local gyms before joining. Look for one that has a commitment to sustainability in terms of energy use, water conservation, water refill points, waste segregation and recycling etc.
- Make sure you place a home-based treadmill or static bike near a window or in a well-ventilated room to minimise the risks of inhaling dust and other air pollutants while exercising.
- Rehome or donate unused gym equipment. Send broken and obsolete equipment for recycling with other electronic waste (see also page 97). Ask your local authority or waste provider where to take such a machine for recycling.
- Choose a gym close to home or work and walk or cycle to add to the fitness regime. Your drive to the gym could be biggest contributor to the carbon footprint of your gym workout.

GYMS IN THE PARK – OPEN FOR ALL

Have you seen an outdoor gym in a park near you? There are more than 1,800 open air gyms in the UK, so you should have access to one.

Many councils over the past few years have been trying to encourage local people to get fit without the need for an expensive gym membership. The result is an outdoor gym comprising a range of specialist equipment designed to stay outside all year, is free for everyone and is easy to use.

Various machines offer low-impact training for the lower body, upper body and core, as well as cardiovascular and resistance exercises. You might see walkers, parallel bars, cross-trainers, hurdles, step-ups and pull-up bars.

The idea of outdoors gyms originated in a national fitness campaign in China before the Beijing Olympics in 2008. Adult gyms or playgrounds were set up in local parks to encourage people to get active, without the cost of joining a gym.

Open-air gyms make exercising accessible to all and are based on nudge theory (that is, you change people's behaviour through making things easier rather than penalising them). So, gyms were located in places people go to anyway, such as beside children's playgrounds.

The Great Outdoor Gym Company in the UK uses 'Cardio Charge technology' on their equipment so that users can charge their mobile devices as they work out – a great incentive to get your heart rate up! They also have an 'Energy Gym' range of equipment such as cross trainers and bikes that produce energy to power the lights at the outdoor gym or nearby buildings, every time they are used.

Energy Gyms can generate 1kWh of energy per gym every day – enough to power a 45W floodlight for twenty-two hours. Some cities, such as San Antonio in Texas, USA, have used health and income data to locate outdoor gyms where people need them most.

Join a fitness group that helps you to get fit in the outdoors, adding the benefits of a regular connection with nature to your mental health.

Additional facts

The vending machine alone at a gym can use about ten times the amount of energy of a home refrigerator.

In 2019, there were over 1,800 outdoor gyms in the parks around the UK – the perfect place for a zero-carbon workout.

A study on a gym in Wake Forest University in the US found that the gym's treadmills and stairmasters alone consume 808kWh of energy each week – that's equivalent to driving 1,397 miles (2,248km) in an average car.

There are over 4 million registered golfers playing on over 7,000 golf courses in Europe – 66 per cent of whom are men, 25 per cent women and 9 per cent juniors.

In total, 16 per cent of the players and 28 per cent of the courses are in England, with 10 per cent of players and 9 per cent of courses in Scotland.

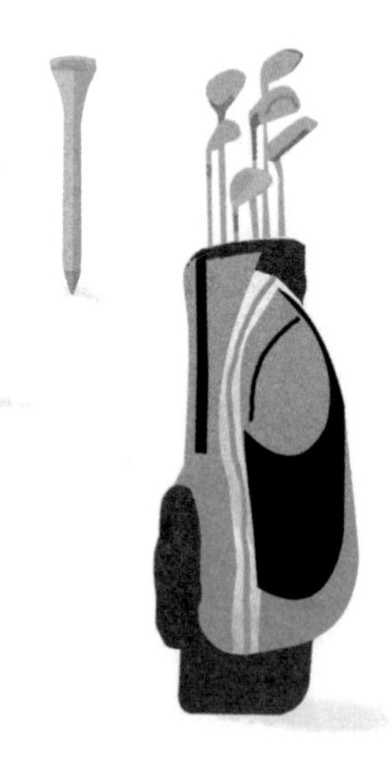

GOLF CLUB

Golf is a much-loved and played sport by many, growing in popularity as it becomes a sport for all rather than as it used to be – an elite activity for Pringle-jumper-wearing businessmen.

There is some controversy over the origins of the game with the Netherlands and Scotland vying for the title of 'the home of golf'. Nevertheless, the Scots can claim the first 18-hole golf course; it was constructed at St Andrews in 1764 with the tees, fairways, roughs and greens typical of golf courses around the world today.

THE IMPACTS

Even though golf is played in the great outdoors in often beautiful locations on the coast, in wooded areas and near lakes and other natural landmarks, it has a bad reputation when it comes to the environment.

First, there's the impact the creation of a golf course has on the natural habitats destroyed to make the artificial landscape, and then there's the amount of water needed to irrigate the grass, as well as the pesticides and herbicides used to keep it looking pristine. It is estimated that a typical golf course needs 378–3,785 litres of water per week in summer in order to remain green and in good condition, with the variability due to climate, soil and vegetation type. But there are measures that can be taken to reduce water use on golf courses – by planting native plants, reducing irrigation, harvesting rainwater and maintaining natural features.

The equipment used for golf – from the clubs and golf balls to tees and golf carts – also has an impact. Golf clubs are made from stainless steel, aluminium, titanium and graphite plus ceramics, wood and synthetic foams. Exotic woods, such as persimmon and

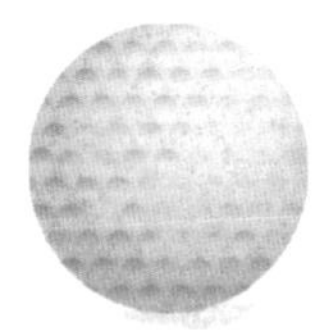

maple, are preserved using oils or polyurethane coatings. The extraction of these raw materials and the energy used to manufacture the drivers or putters all adds to the carbon footprint of the golf clubs.

Old clubs are not easily recyclable because they are made of many different materials, but clubs made predominantly from metal should be of interest to a scrap metal dealer.

Golf balls and golf tees often contribute to litter as they get lost and left behind. It is estimated that 300 million golf balls get lost or are thrown away every year. Golf balls are often made from heavy metals, such as tungsten, cobalt and lead, with a rubber core, all wrapped up within a hard plastic coating.

While a biodegradable ball has not yet been invented, and I don't think any golfer wants to go back to the early days of wooden balls, greener options made from bamboo and cotton that are not chemically treated and that use alternatives to heavy metals to weight the ball are now available.

Left in the soil, plastic tees cause damage to lawnmowers and break down slowly into microplastics that enter the soil and eventually the groundwater. Cheaper plastic tees break easily and typically last for a mere four or five rounds; alternatives are available in the form of wood and bamboos tees as well as versions made from recycled plastics. Makers of tees from recycled plastic claim they are the most durable, meaning you will need fewer of them, thus reducing their environmental footprint. That said, wood and bamboo tees are also biodegradable.

Golf carts are another component of the golf game. Most are electric and don't use petrol or diesel and the positive impact can be multiplied by making sure that golf courses produce their own electricity or buy in from a renewable energy supplier. There are now solar panels to install on top of golf carts, so that they run on solar energy, and solar-charging points for electric golf carts exist too.

REHOMING UNUSED CLUBS

Garage2green, aptly based in Scotland, aims to get old and unused golf clubs out of people's sheds and garages and into use, enabling new golfers to join the sport. Supported by Zero Waste Scotland, the aim is to play a part in reducing waste by collecting, repairing and reusing golf clubs. They are also looking for opportunities to remake old clubs into new products.

WHAT YOU CAN DO

- Encourage your golf club to be more environmentally friendly by installing water-saving strategies, by planting indigenous flowers and plants to create habitats for biodiversity, by not using pesticides and weedkillers, by generating or buying renewable energy and by reducing waste.
- Look after your clubs. Learn how to repair them from fellow players, your golf pro or online videos.
- Research before you buy new golf balls, tees and clubs and find the most sustainable options available to you.
- Pass on old and unwanted golf clubs to friends, to local kids who want to try golf, to your local mini-golf set-up or a charity when it's time to upgrade your clubs.

After 2018's London marathon, Westminster City Council collected 5,200kg of rubbish and about 47,000 plastic bottles for recycling from the streets. In 2019's event, over 350,000 plastic bottles were collected and recycled because there was a substantial effort made to clear up and make the event greener.

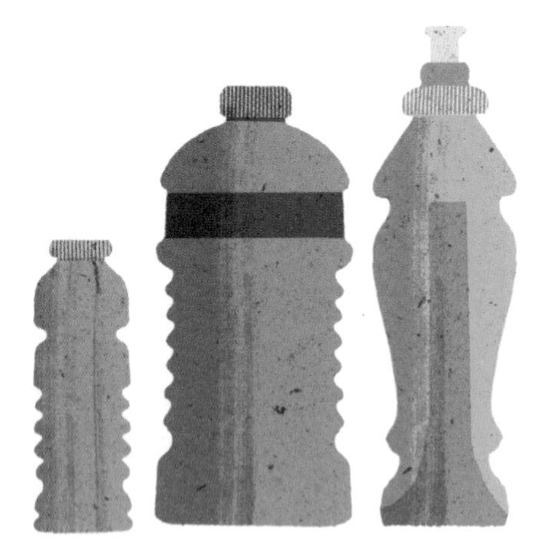

SPORTS DRINK AND FOOD

The rise in popularity of marathons and ultra-marathons, triathlons and Ironman events, extreme swimming and ultra-distance cycling have all contributed to the growth of the specialised sports drink and food industry. Sports gels, isotonic, hypertonic and hypotonic drinks as well as protein powders line the shelves of sport stores.

Experts recommend sports gels, isotonic drinks and protein shakes for athletes doing intensive exercise for over an hour to replace water and electrolytes lost through perspiration and to restore carbohydrates and keep energy levels up. But, if you are going for a run for less than an hour, doing a 45-minute fitness class or swimming your usual fifty lengths of the pool, then water is all you need.

THE IMPACTS

Sports drinks and gels are made from fluids, electrolytes and carbohydrates including sugars. Research by the University of Oxford has questioned the claims of some sports drinks in terms of their ability to enhance performance, saying the research the claims are based on is inadequate. The risks to children of obesity and tooth decay associated with consuming such high-sugar sports drinks has been highlighted by the *British Medical Journal,* which recommends these are only used by athletes.

Sports drinks generally come in plastic bottles, while gels come in plastic pouches. The bottles can be recycled if they are empty and clean, but the pouches are not currently recyclable. As sports gels and drinks are used 'on the move' in races and during training, there is a good chance that they'll be dropped on the ground and simply become litter.

Race organisers typically provide bottles of water and/or sports drinks for competitors to keep hydrated during the event, but most bottles grabbed from such drink stations are only partially used and then cast aside to be picked up by volunteers later.

The pouches that gels come in have a tab that you have to tear off before you squeeze out its contents. Unfortunately, while athletes might do their best to make sure empty pouches get into a bin, the tabs are so small they get lost along the way, ending up in ditches, streams and eventually the sea. Designing a tab that stays attached to the pouch after opening would be a really good green idea!

ECO-FRIENDLY SPORTS DRINKS AND FOOD

Sports gels, drinks and water are starting to become available in edible capsules made from seaweed and other plants. Oohos are edible water capsules that were invented by a London-based company and can be made on location at your event to reduce the carbon footprint associated with transport. You simply pop the whole thing in your mouth – so it's zero waste.

Gone is a 100 per cent biodegradable plant-based packaging for sports drinks and gels that breaks down in a number of days and is intended for use in competitive events.

WHAT YOU CAN DO

- Use reusable plastic cups and water bladders, like ultra-marathon runners do – you can refill at water fountains if you're training or racing. Other sports could adopt a similar approach to cut their environmental impact.
- Investigate alternative drink containers. Some marathons use recyclable or compostable drinks cups as an alternative to plastic bottles; others opt for a collapsible pouch with a spout that makes it easier to drink out of than a cup – these containers are designed to be held in your hand or hooked into a belt or bag.
- Make your own sports gels and drinks and use reusable containers. Both carbohydrates and electrolytes come in powder and dissolvable tablet form and buying in bulk saves money and reduces waste.
- Take litter home. If you use gels during competitions and training, put the empty package in your shorts or even inside your swimsuit until you get to a rubbish bin.
- Get a good-quality reusable bottle and refill from a tap or water fountain – most of us only need water for our workouts.

Additional fact

Croyde Ocean Events and Plastic Free North Devon, in association with The Pickwell Foundation, have developed a plastic-free sports event toolkit on how to design a sustainable sporting event.

Broken umbrellas create over 150,000 tonnes of wasted metal each year (enough to build twenty-five Eiffel Towers).

The city of Songxia in China produces half a billion umbrellas per year, which accounts for only 30 per cent of the Chinese umbrella market, and employs 40,000 people.

UMBRELLA

Umbrellas – you either love them or hate them. While some people wouldn't go out without one, others hate them poking in their eyes and blowing inside out with the slightest gust of wind. They are of varying quality and some have almost become disposable. Many of the cheaper models break easily and end up stuffed into bins on rainy days.

The umbrella has a pretty universal design – it hasn't changed much in 3,000 years. The first umbrellas were used in Egypt and China to protect people from the sun, often held by servants and adding to the status of well-to-do members of society.

Umbrellas didn't catch on in Europe until the late 16th century when they were used by clergy to signal their status in society, and became a fashion item for women in the form of a sun parasol in the 18th and 19th century.

Englishman Jonas Hanway made the first rain umbrella in the mid-1700s and Samuel Fox from Sheffield modified the design to use a lighter, steel frame (before that the frame had been wooden) in 1852. Nylon and polyester became the fabrics of choice for umbrellas from the 1960s and over time umbrellas appeared in all colours and sizes, from golf umbrellas to foldaway pocket versions.

THE IMPACTS

Modern umbrellas have steel frames, nylon or polyester canopies and plastic or wooden handles. Steel-making requires the extraction of iron ore and the use of energy to smelt the steel. Nylon, as with all synthetics, is made from petrochemicals and the manufacturing process releases CO_2 and nitrous oxide (a very powerful greenhouse gas). Then, add the energy needed to make the umbrella and ship it from China, where most of them are made, to your local retailer and the carbon footprint of an umbrella jumps up again.

Finally, add what's known as the end-of-life phase of the umbrella – that is, when it breaks and becomes waste, as rubbish either in a landfill or in an incinerator. The worst-case scenario for a cheap fold-up umbrella might be to be used for only one day before it is lost or broken – that's a lot of energy, resources and waste for a brief period of protection from the wet stuff.

As if all that wasn't enough, umbrellas usually come in a small nylon pocket, to keep it nice and neat and dry in your bag. Most people lose these the first time they use the umbrella. They escape from pockets, get left behind and are fiddly to get back on an umbrella after you use it, so they are cast aside. Then, to make things worse, some shops and hotels provide disposable bags to put your wet umbrella in to stop it dripping on the floor – yet more single-use plastic bags that can't be recycled.

Most umbrellas are not designed to be repaired, so whether you get a good one and look after it carefully, or put up with a cheap, essentially disposable option is up to personal choice. Also, because an umbrella comprises many different materials, they are not recyclable. A new approach to umbrella design is well overdue.

Designers are currently working on reducing the number of materials and parts in umbrellas to make them more durable, easier to repair and recyclable. Some companies have also started to make umbrellas from recycled materials, using fabric made from recycled plastic water bottles for the canopy, bamboo or FSC-certified wood for the handles and recycled aluminium or bamboo for the shaft and frame.

A REPAIRABLE AND RECYCLABLE UMBRELLA

UK designer Ayca Dundar has made an umbrella called the Drop from just six parts after being horrified at the numbers of broken umbrellas abandoned on London's streets. Her design is repairable and works well on windy days.

Her Ginkgo umbrella, designed in Italy, is made from twenty pieces of just one material and is completely recyclable.

WHAT YOU CAN DO

- Wear a coat with a hood and avoid umbrellas completely.
- Buy the best-quality umbrella you can afford and take care of it. Choose one without a cover and ideally from recycled or sustainable materials.
- Say 'no' to umbrella covers offered by shops and hotels on wet days.
- Don't accept cheap umbrellas as promotional giveaways.

Additional fact

Research conducted in hospitals in Victoria, Australia in 2012 found an average of twenty umbrella-related injuries a year across thirty-eight accident and emergency wards.

Research in the UK found that 51 per cent of consumers buy 'fast fashion' sunglasses from online and high-street fashion shops rather than an optometrist.

SUNGLASSES

Sunglasses have two functions: to protect your eyes and to look stylish. The common emphasis on style over protection leads people to own multiple pairs of sunglasses and to an increased demand for cheaper, more disposable glasses over more expensive ones.

Protecting your eyes from sunlight is as important as protecting your skin because exposing your eyes to UVA and UVB rays increases the risk of eye conditions such as cataracts and macular degeneration. The World Health Organization estimates that up to 20 per cent of cataracts may be caused by overexposure to UV radiation, meaning they could be avoided.

THE IMPACTS

Sunglasses are made of a mix of plastic, metal and glass, which makes them difficult to recycle. Their production involves the extraction of natural resources, the use of chemicals and the consumption of energy. As a result, a pair of glasses has a carbon footprint of approximately 4.8kg of CO2e, equivalent to charging a smartphone 607 times. Given that some people own several pairs of sunglasses at any one time the impact soon adds up even before you consider that broken and discarded sunglasses usually end up in landfill or are incinerated.

Sunglasses that are made to last and that can be repaired can last ten years or more. Think of the current demand for retro and vintage sunglasses – if you'd held on to your old ones from the 1980s they would be ready to be worn again now!

The biggest offenders are cheap sunglasses as they are both unlikely to protect your eyes and are liable to break or scratch easily, making them largely disposable.

Lightweight lenses are made from CR39 (a plastic resin) or polycarbonate. Both are plastic and neither biodegrades. Recycled plastic has started to be used by environmentally conscious brands to make sunglasses, as well as materials such as bamboo, wood and plant-based acetates.

Dick Moby, a company that makes sustainable sunglasses, estimates that a pair of their recycled acetate glasses saves 2.56 litres of water compared with a pair of glasses made from virgin acetate and that their recycled stainless-steel frames produce just 0.05kg CO_2e compared with 0.12kg for glasses made from stainless steel.

WHAT YOU CAN DO

- Look after the sunglasses you have and keep them in a box to prevent them getting scratched or damaged.
- Try having your sunglasses repaired. If you have a pair of sunglasses with scratched lenses but the frames are in good condition, an optician may be able to restore or replace the lenses. Sunglass Fix based in Australia but operating worldwide sells replacement lenses.
- Think before you buy another pair. How many pairs of sunglasses do you really need? Look out for sunglasses made from ghost fishing nets, such as those made by British company Waterhaul, who also offers a lifetime guarantee that they will take back your glasses to fix them or recycle them. Steer clear of 'buy one get one free' offers unless you really do need that second pair.
- Hold on to sunglasses that are good quality but go out of fashion or aren't quite your style. They are more than likely to come back into fashion one day and you will be delighted to have them – if not, pass them on to a friend or sell them.
- Look for brands that use recycled and sustainable materials in their frames and that provide parts and support to fix your glasses and extend their life, and ask if they offer a 'take back' service to recycle old glasses.

CONTACT LENSES

In the UK, 20 per cent of contact lens wearers admit to disposing of their lenses down the toilet or the sink. Never flush your disposable contact lenses down the toilet, they are made of plastic and pose a threat to wildlife if they enter the waterways.

Contact lenses come in blister packs, which are not commonly recyclable. However, Bausch + Lomb have set up a recycling programme for their blister packs with recycling organisation TerraCycle. You can now post the empty blister packs to TerraCycle or drop them off at a participating optometrist.

The UK's first free national recycling scheme for plastic contact lenses was rolled out in January 2019. Wearers of any brand of soft lens can have their lenses and packaging collected or drop them off at recycling bins at Boots Opticians and selected independent stores. The recycled contact lenses, blister and foil packaging will be turned into products such as outdoor furniture.

40 per cent of adults in the UK read printed newspapers in 2013. By 2016 this had fallen to 29 per cent, largely due to the rise of internet-based news.

NEWSPAPER

How do you like to read your news? Via twitter, a broadsheet, online or a tabloid newspaper? The choice we have today is greater than ever before and that poses challenges to the traditional printed newspaper.

The earliest newspapers date back to the start of the 17th century. Before that, announcements related to specific events were sometimes issued as manuscripts or hailed by a town crier.

In fact, the Japanese, who have a long tradition of newspapers to this day, have published news since 1603. The top two most-read newspapers today are both Japanese – the *Yomiuri Shimbun* and the *Asahi Shimbun*.

The first US newspaper had a great name – *Publick Occurrences Both Forreign and Domestick* – and was issued only once, in September 1690, before it was banned by a colonial governor.

In the UK newspapers with long histories include *The Times* founded in 1785 and *The Observer* in 1791.

Printed newspapers continue to be published and sold today but are gradually being overtaken by online news and social media feeds. *The New York Times*' executive editor, Dean Baquet, has predicted that local newspapers will be gone by 2024. In the UK, 245 local newspapers closed between 2005 and 2018 and with them the loss of hundreds of jobs.

THE IMPACTS

Printed newspapers have various impacts on the environment. Trees are cut down to make paper, the pulping of the paper uses lots of water and chemicals and the printing and the transport of the paper add to its carbon emissions (see also pages 193–195). All that comes before the disposal or recycling of the paper once its day is up. It is estimated that 75,000 trees are used to print the Sunday edition of *The New York Times*.

In his book, *How Bad are Bananas?*, Mike Berners-Lee calculated the carbon footprint of different newspapers. He found that a *Guardian* newspaper that is recycled emitted 0.8kg CO2e, while a weekend paper with all of its supplements racked up 1.8kg CO2e if recycled compared with 4.1kg CO2e if it went in the bin. So, binning your paper rather than recycling more than doubles its carbon footprint.

So, is it better to read your news as a paper, on your laptop or a tablet? A study published in 2013 compared the environmental impact of reading a printed paper to reading a paper on a computer or an e-reader. The study found that if you live in Europe (where some electricity is from renewable sources) and you spend minutes a day reading the news, then reading online or with an e-reader produces less carbon emissions than a printed paper. But, if you read for more than 30 minutes, the print version of the paper has a lower carbon footprint (28kg CO2e/person/year for printed paper versus 35kg CO2e/person/year for online reading).

COMPOSTABLE PLASTIC WRAP

The weekly and Saturday editions of *The Guardian* switched from a plastic wrap to a compostable wrap made from potato starch for their supplements in 2019.

Be sure to put this type of wrap in your food or garden waste bin.

In the case of printed newspapers, the making and printing of the paper are significant contributors to the carbon footprint, while for online reading it is the energy required to power the device that is most important (even more so than the energy and materials required for the manufacture of the product). As more energy switches to renewable sources, the carbon footprint of reading online will continue to drop.

WHAT YOU CAN DO

- Get your news in a way that suits you. If you read a printed paper slowly over a few days or the course of a week and then reuse it as cat litter, as a substitute for firelighters or recycle it – that is OK. If you mostly use your phone to follow news, reading most things for only a few minutes, then stick with your phone or online news. For online news make sure you keep your electronic devices for as long as possible to reduce waste – if they start to slow down, clear out the hard drive or get them refurbished. Avoiding waste is key to reducing your footprint.
- Switch to an energy supplier of renewable electricity to reduce the carbon footprint of any online reading.
- Avoid newspapers with lots of supplements in plastic bags that can't be recycled. If you do get them, read them and then pass on all the supplements for others to read, perhaps at a local dentist or doctor's surgery.
- Support your local newspaper – local jobs and a thriving community are essential for sustainable societies.

Additional fact

There are only 300 newsstands in New York City today compared with 1,525 in the 1950s when morning and evening editions of daily papers were on sale.

About 2.2 million books are published each year. China and the US top the list of publishers worldwide, but the UK publishes more books per capita than any other country.

BOOK

There is no greater pleasure than escaping within the pages of a good book. Whether packing for holidays or taking a trip, many of us will be sure to have a book or an e-reader in our bag.

Our fascination with books dates back to 868AD when the first book was printed in China. By the end of the 15th century more than 20 million books had been printed across Europe and we have been making books ever since.

The first Penguin paperback was published in 1935, making books affordable so that they could be read by everyone. The next big development was the arrival of e-books in the early 2000s and the invention of the e-reader (the Amazon Kindle) in 2007 that can hold hundreds of books in one place.

Audio books are another, increasingly popular way of accessing books (fiction and non-fiction). Making books accessible to those with vision problems, audio books also offer people a way to listen to books as they commute.

THE IMPACTS

The manufacture of printed books uses paper and so all the environmental impacts of paper-making apply here too (see pages 193–195).

As publishers make more efforts to print on recycled, FSC-certified and chlorine-free paper, to use vegetable dyes and to reduce the weight of books, their environmental footprint decreases. This book is printed cover to cover on FSC-certified paper, with vegetable-based ink and no plastic lamination on the paper cover.

E-readers have grown in popularity but they have an environmental impact, too – including the raw material required to make them, the energy required to power them and the footprint they create as e-waste. So, which is better – a paper book or an e-book?

I'm afraid there is no simple answer. By one estimate the energy, water and raw materials needed to make an e-reader is equivalent to forty or fifty books and the carbon emissions of using an e-reader are roughly equivalent to 100 books.

The emissions associated with an e-reader are largely associated with how much you use it, because the energy associated with powering it up is a significant part of its carbon footprint. Also whenever you upgrade your e-reader you add to the carbon footprint. For a paper book, once it's made the carbon footprint doesn't change unless you start mailing it to people around the world!

Audio books are comparable to e-readers in that their environmental impact is primarily down to the energy used to power the device you use to listen to them, but they do have the advantage that you can listen to them on your phone, so you don't need an additional device to access them.

So, if you read more than 100 books on your e-reader, before you change it, it is likely to be more environmentally friendly than buying 100+ new books. That said, if you read fewer than 100 books in the same time period, you are probably better off buying the printed book.

In reality, however, many of us do both, and so the comparison of one versus the other becomes pointless. In addition, many of us prefer one way of reading over the other and the decision between book or e-reader may even affect our ability to comprehend or retain the information we are reading – so it is a highly personal decision. But as a rule, printing books that don't get read just contributes to waste, but printing books that are read and shared and that improve and influence people's lives makes them treasured possessions.

WHAT YOU CAN DO

- Share your printed book once you have read it and pass it on – it's the most environmentally friendly thing you can do. If you find you are buying books and not reading them, try to buy fewer.
- Join a library and borrow books – that's super-sustainable.
- Buy second-hand books instead of new books when you can – try to buy from a local or national reseller to cut down on the carbon associated with any transport.
- Recycle damaged books that are at the end of their life, so that they can become paper again and avoid the emissions associated with going to landfill.
- Keep an e-reader as long as you can and read lots of books to make the initial investment and its associated emissions worthwhile.
- Power your e-reader and the device you use to listen to audio books with renewable energy to reduce the carbon footprint of its use and make sure it gets recycled with electronic waste or e-waste at the end of its life (see also page 97).

Additional facts

A 2014 survey in the UK revealed that:

- *56 per cent of the population read printed books.*
- *23 per cent of readers use an e-reader and read printed books.*
- *11 per cent of UK readers mostly use an e-reader.*

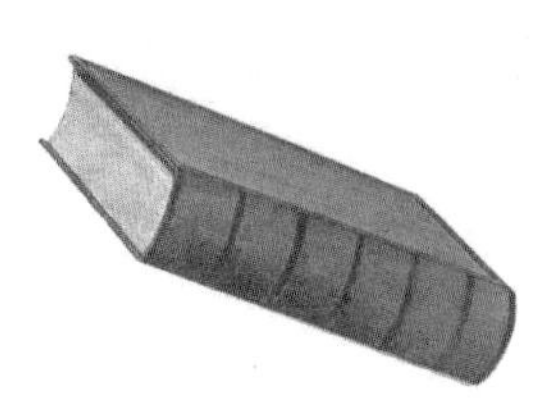

PARTIES AND CELEBRATIONS

500 million straws are used every day in the US – enough to wrap around the Earth four times.

Approximately 4.7 billion plastic straws are used in England each year.

Ten million straws are used every day in Australia – end to end that is the same distance as from Cairns to Melbourne.

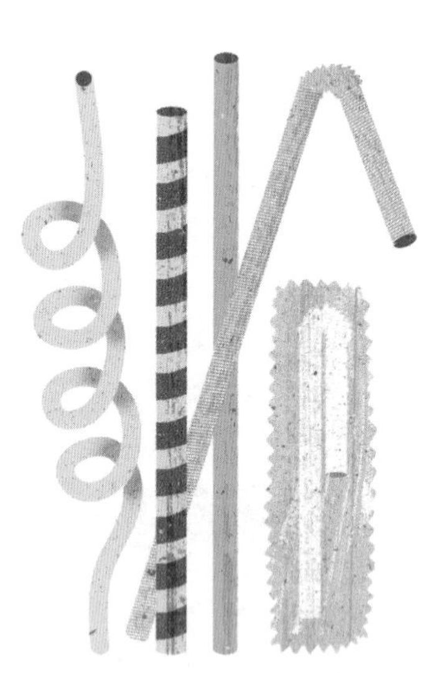

STRAW

Plastic straws have been in the news for all the wrong reasons in recent years. A video of a marine biologist removing a plastic straw from the nostril of a sea turtle in 2015 went viral, causing a massive emotive reaction in the public sphere. People who saw it questioned the need for straws when they could cause such pain and suffering to a marine creature.

People have been using straws for thousands of years with evidence that brewers in Mesopotamia used metal straws to sample beer from fermentation jars without getting all the bits settled on the top.

The first patent for a drinking straw was lodged in 1888 by a man called Marvin Stone who sought a durable replacement for a rye grass straw. He made the prototype paper straw and started manufacturing them in 1890.

Bendy straws came along in the 1930s, invented by Joseph Freidman, and later called the Flex-Straw. Straws switched from paper to plastic in the 1960s and never looked back. Plastic ones were cheaper and lasted longer than paper straws and could be produced in different sizes, shapes and colours.

Straws play an important role in the lives of people with medical conditions or disabilities who are unable to drink from a cup, but their use as a single-use, throwaway item has come under scrutiny and has led the European Union to ban their use from 2021.

THE IMPACTS

Plastic straws are made from polypropylene, colours and other additives, such as plasticisers and antioxidants, that prevent interactions between the plastic and oxygen, to make the straw more durable. Polypropylene is made of petroleum and requires energy to be made into a straw, both of which contribute to the carbon footprint of the finished product. Added to that is their packaging and transport.

Plastic straws are not recyclable. Although polypropylene is technically recyclable, the facilities don't currently exist to recycle straws. It takes 200 years or more for a plastic straw to break down, and even then they persist as microplastics, which can be eaten by wildlife if they escape into the environment.

Plastic straws are among the ten most frequently found items of litter in coastal clean-ups conducted by the Ocean Conservancy in the US and in the top thirteen single-use plastic items found on beaches across the EU. A study in 2017 estimates that as many as 8.3 billion plastic straws pollute beaches all around the world.

WHAT YOU CAN DO

- Avoid plastic straws. Refuse them if offered in a bar, café or takeaway. Most people are able to enjoy their drink without a straw.
- Ask your local café, pub or bar to stop using plastic straws. Use some facts from this book to convince them. Give your support to businesses that no longer use plastic straws.
- Join campaigns around the world to end the use of straws in schools. School milk is still delivered in Ireland and in the UK in cartons with plastic straws wrapped in a plastic film. Neither the straws nor the film are recyclable and both cause litter. Plastic straws will be banned throughout the EU, and most likely in the UK as well, by 2021 but campaigns can make the change happen before then. In Australia the Straw no More campaign is working with councils to end the use of straws in schools.
- Look for alternatives. Paper straws are a single-use alternative but really single-use anything is a waste of resources, so try a reusable option instead. Reusable straws are made out of stainless steel, bamboo and even glass. You can even buy a folding metal straw that comes in a case, so that you always have a straw to hand, if needed.

WHAT'S WITH STIRRERS?

It can be so disappointing to order a well-earned gin and tonic or a cocktail only to see it arrive to the table with not just a plastic straw but also a plastic stirrer.

Plastic stirrers are far from flimsy – they are made from high-quality plastic and could last for hundreds of years, rather than the two minutes or so you will use it for.

So, do we need stirrers? Most people stir once and then take the stick out of the way. For those that must stir – look out for reusable metal stirrers or combined straw/stirrers made from stainless steel. Nowadays you can even get compostable stirrers made from wood or bamboo.

Additional fact

Plastic and other types of marine litter are concentrated in parts of the oceans in large rotating currents called gyres. There are five ocean gyres and the largest one – the Great Pacific Garbage Patch (see also page 134) – occupies an area twice the size of Texas.

EDIBLE STRAWS

Sustainability is all about innovation and finding new ways to do things that have less impact on the planet. A great example of this is the company Loliware.

Chelsea Briganti and Leigh Ann Tucker are both industrial designers in the US. When faced with the reality of the impact of plastic cups and straws on wildlife and the environment, they decided there had to be a different way of doing things. They wanted to move past incremental change to radical change in design to see what could be achieved – they wanted to make single-use items that would disappear.

So, they invented a hyper-compostable edible straw called the Lolistraw. They re-engineered the straw through a process they call 'joyful innovation', through which they make design and problem-solving fun, and came up with a product that is good for people and for the planet.

The straws are made from seaweed and look and perform just like plastic but are instead made from food-grade materials. So, when you finish using your straw you can eat it! The straws can be used in liquid for up to 18 hours, so they perform much better than paper straws, and break down in a marine or natural environment – although the aim is for them never to reach the natural environment as they should be eaten or composted first. Their colours come from fruit and vegetable dyes, meaning they are completely safe to eat.

Chelsea and Leigh Ann have used crowdfunding to commercialise their invention – which they expect to be on sale from the end of 2019.

The average child owns 238 toys but plays with their twelve favourites most of the time, meaning that 95 per cent of toys are rarely used.

TOYS

If you have kids, then toys are part of daily life – you trip over them, tidy them away and protect the favourites for fear of loss and heartbreak.

The earliest toys were animal-like toys, with balls, kites and yo-yos following afterwards. While it's apparent that paper, wood, clay and string were the materials used in early toys, metal and plastic are more common in modern times. Nowadays, a staggering 90 per cent of toys on the market are made from plastic.

The overpurchasing of toys is linked to the general trends in consumption globally, where we just keep buying ever more stuff, much of it on credit. The average American household, for instance, has over $15,000 in credit card debt and generates 230 metric tonnes of rubbish a year. Families in the UK don't come off much better – with an average household consumer debt of £6,454 and 91 metric tonnes of waste.

Toys play an important role in a child's development and education, but many toys contribute little to learning while creating a lot of waste. So, it makes no logical sense to have hundreds of toys for every single developmental stage of childhood.

TOY SPENDS WORLDWIDE

The UK spends over £3 billion a year on toys. The average spend in 2017 was £339 a child.

In Ireland, a 2015 study showed that the average Irish parent spends €254 on presents per child at Christmas, while 16 per cent could spend up to €600 on each child.

Australian parents are the biggest spenders, handing over the equivalent of £555 per child on toys and games in 2013.